SpringerBriefs in Molecular Science

W0037345

For further volumes:
http://www.springer.com/series/8898

Igor Ying Zhang · Xin Xu

A New-Generation Density Functional

Towards Chemical Accuracy for Chemistry of Main Group Elements

 Springer

Igor Ying Zhang
Xin Xu
Department of Chemistry
Fudan University
Shanghai
People's Republic of China

ISSN 2191-5407   ISSN 2191-5415 (electronic)
ISBN 978-3-642-40420-7   ISBN 978-3-642-40421-4 (eBook)
DOI 10.1007/978-3-642-40421-4
Springer Heidelberg New York Dordrecht London

Library of Congress Control Number: 2013947359

Printed on acid-free paper

Springer is part of Springer Science+Business Media (www.springer.com)

Preface

Density functional theory (DFT) has revolutionized the role of theory by providing accurate first principles predictions of critical properties for applications in physics, chemistry, biology, and materials science. Despite dramatic successes, there remain some serious deficiencies in describing, for example, weak interactions (London dispersion) which are so important to the binding of drug molecules to proteins, and reaction barrier heights which are so important to the understanding of chemical reactions. This book covers the most recent progress in this area by presenting a new generation of DFT that dramatically improves the accuracy for these properties by including the role of the virtual (unoccupied) states. The underlying physics of this so-called doubly hybrid density functional (DHDF) is explored and the performance in the description of thermochemistry, thermochemical kinetics, and nonbonded interactions is demonstrated using some well-established benchmarking data sets. Here we consider only finite systems (atoms, molecules, etc.). The present book shall be of interest to college students majoring in physical chemistry as well as to all computational chemists. The authors are glad to receive any helpful suggestions and comments from the readers.

We are deeply grateful to the group members and our family members for their continuous supports. We acknowledge the financial supports from the National Natural Science Foundation of China, the Ministry of Science and Technology of China, and Fudan University.

Shanghai, April 15, 2013

Igor Ying Zhang
Xin Xu

Contents

Chapter 1
An Overview of Modern Density Functional Theory

Abstract Density functional theory (DFT) has become the leading method in computing the electronic structures and properties from first principles. Its foundation was laid on Hohenberg-Kohn theorems, which proved that there exists a one-to-one correspondence between the ground state electron density ρ_0 of a many-body system and its total energy. In practice, DFT is most frequently applied in the framework of Kohn–Sham (KS) scheme, where an approximate exchange-correlation functional has to be chosen. Hence, the success of a DFT calculation critically depends on the quality of the exchange-correlation functional. In this chapter, we first briefly discuss the Hohenberg-Kohn theorems (Sect. 1.1). After introducing the KS scheme, various approximations for the exchange-correlation functionals are presented in Sect. 1.2. These functionals are grouped according to Perdew's classification of Jacob's ladder. Finally, some general trends for the functional performances along the Jacob's ladder are outlined.

Keywords Density functional theory · Hohenberg-Kohn theorems · Kohn–Sham scheme · Exchange-correlation · Jacob's ladder

1.1 Foundation of Modern DFT

1.1.1 Wavefunction Versus Density as Basic Variables

The ultimate goal of most quantum chemical applications is to solve the time-independent, non-relativistic Schrödinger equation: [1, 2].

$$\hat{H}\Psi(\vec{r}_1\sigma_1,\ldots,\vec{r}_N\sigma_N) = E\Psi(\vec{r}_1\sigma_1,\ldots,\vec{r}_N\sigma_N) \tag{1.1}$$

where $\hat{H}$ is the Hamilton operator for a molecular system consisting of N electrons in the presence of an external potential $v_{\text{ext}}(\vec{r})$.

I. Y. Zhang and X. Xu, *A New-Generation Density Functional*,
SpringerBriefs in Molecular Science, DOI: 10.1007/978-3-642-40421-4_1,
© The Author(s) 2014

$$\hat{H} = -\frac{1}{2}\sum_{i=1}^{N}\nabla_i^2 + \sum_{i=1}^{N}v_{ext}(\vec{r}_i) + \sum_{i=1}^{N}\sum_{j>i}^{N}\frac{1}{|\vec{r}_i - \vec{r}_j|} \tag{1.2}$$
$$= \hat{T} + \hat{V}_{ext} + \hat{V}_{ee}$$

The first term on the right describes the kinetic operator $\hat{T}$, where the Laplacian operator ∇_i^2 is defined as a sum of second-order differential operators (in Cartesian coordinates)

$$\nabla_i^2 = \frac{\partial^2}{\partial x_i^2} + \frac{\partial^2}{\partial y_i^2} + \frac{\partial^2}{\partial z_i^2} \tag{1.3}$$

The last term in Eq. 1.2 describes the electron–electron repulsion $\hat{V}_{ee}$ that sums over all distinct pairs of different electrons. The external potential $v_{ext}(\vec{r})$ generally comes from the electrostatic interaction between electrons and nuclei, where $\{Z_\alpha, \vec{R}_\alpha\}$ indicate the nuclear charges and their spatial coordinates, respectively.

$$v_{ext}(\vec{r}) = -\sum_{\alpha}\frac{Z_\alpha}{|\vec{R}_\alpha - \vec{r}|} \tag{1.4}$$

$\Psi(\vec{r}_1\sigma_1, \ldots, \vec{r}_N\sigma_N)$ stands for the wavefunction of the system, which depends on the $3N$ spatial coordinates $\{\vec{r}_i\}$, and the N spin coordinates $\{\sigma_i\}$ of the electrons. The wavefunction Ψ contains all information that can possibly be known for the quantum system that we are interested in. Nevertheless, as electrons are interconnected by $\hat{V}_{ee}$, it is very difficult to find a good approximation to the solution of Eq. 1.1 for $N > 1$ [1–4].

The electron density $\rho(\vec{r})$ is given by [1–4]

$$\rho(\vec{r}) = N\int \ldots \int |\Psi(\vec{r}\sigma_1, \ldots, \vec{r}_N\sigma_N)|^2 d\vec{r}_2 \ldots d\vec{r}_N d\sigma_1 \ldots d\sigma_N \tag{1.5}$$

It determines the probability of finding any of the N electrons within the volume element $d\vec{r}$ but with arbitrary spin while the other $N - 1$ electrons have arbitrary positions and spins in the state represented by Ψ. As Ψ is normalized, it is clear, from Eq. 1.5, one has

$$\int \rho(\vec{r})d\vec{r} = N \tag{1.6}$$

Unlike the wavefunction, electron density always depends on only three spatial coordinates, and it is an observable that may be measured experimentally via, e.g., X-ray diffraction. If, instead of the complicated N-electron wavefunction Ψ, the electron density ρ could be used as the basic variable, such an approach would be physically more appealing and computationally more efficient.

1.1.2 Hohenberg-Kohn Theorems

Hohenberg-Kohn theorems put the modern density functional theory (DFT) onto a firm physical foundation [2–6]. The first Hohenberg-Kohn theorem is often quoted as "proof of existence" [5], which stated that "the external potential $v_{ext}(\vec{r})$ is (to within a constant) a unique functional of $\rho(\vec{r})$; since, in turn $v_{ext}(\vec{r})$ fixes $\hat{H}$ we see that the full many particle ground state is a unique functional of $\rho(\vec{r})$" [5]. The existence of a one-to-one correspondence between the ground state electron density ρ_0 of a many-body system and its Hamiltonian $\hat{H}$ equivalently proved that there exists an energy functional in terms of density, that is $E[\rho]$.

At this point, it is instructive to separate the energy expression of $E[\rho]$ into its components, as the external potential energy $V_{ext}[\rho]$, in terms of electrons-nuclei attractions, is readily expressed as a functional of ρ.

$$\begin{aligned} E[\rho] &= F_{HK}[\rho] + V_{ext}[\rho] \\ &= F_{HK}[\rho] + \int \rho(\vec{r})v_{ext}(\vec{r})\mathrm{d}\vec{r} \end{aligned} \quad (1.7)$$

$F_{HK}[\rho]$ is the so-called Hohenberg-Kohn functional, which shall consist of the kinetic energy functional $T[\rho]$ for electron and the electron–electron repulsion energy functional $V_{ee}[\rho]$.

$$F_{HK}[\rho] = T[\rho] + V_{ee}[\rho] \quad (1.8)$$

As the classic Coulomb energy $J[\rho]$ among electrons is the major contributor of $V_{ee}[\rho]$, and its expression as a functional of ρ is known, it is also instructive to separate $J[\rho]$ from its non-classic counterpart $E_{ncl}[\rho] = V_{ee}[\rho] - J[\rho]$.

$$J[\rho] = \frac{1}{2} \iint \frac{\rho(\vec{r}')\rho(\vec{r})}{|\vec{r}' - \vec{r}|} \mathrm{d}\vec{r}' \mathrm{d}\vec{r} \quad (1.9)$$

Hence the energy expression for $F_{HK}[\rho]$ now reads as

$$F_{HK}[\rho] = T[\rho] + J[\rho] + E_{ncl}[\rho] \quad (1.10)$$

As every electron is identical, $F_{HK}[\rho]$ is a universal functional, whose form is independent of a particular molecular system (i.e., independent of charges and positions of the nuclei).

The second Hohenberg-Kohn theorem is the variational principle in terms of density [5]. It tells us that functional $E[\rho]$ attains its minimum value with respect to all allowed densities (i.e., Eq. 1.6 is fulfilled), if and only if the input density is the true ground state density.

$$E[\rho_0] = \min_{\rho \to N} E[\rho] \quad (1.11)$$

This constrained variation on density is equivalent to solving the Euler Equation 1.13 [2–6].

$$\delta\left[E[\rho] - \mu \int \rho(\vec{r})\mathrm{d}\vec{r}\right] = 0 \tag{1.12}$$

$$\mu = v_{ext}(\vec{r}) + \frac{\delta F_{HK}[\rho]}{\delta \rho(\vec{r})} \tag{1.13}$$

Hence the electronic chemical potential appears as a Lagrange multiplier μ in DFT.

The Hohenberg-Kohn theorems are the bedrocks of modern DFT, which prove the unique mapping between the ground state density $\rho_0(\vec{r})$ and the ground state energy $E[\rho_0]$. They, however, do not provide any guidance on how $F_{HK}[\rho]$ should be constructed for practical use. In fact, $T[\rho]$ is a major contributor of $F_{HK}[\rho]$. Even the 1 % error in the kinetic energy will prevent DFT from being used as a quantitative predictive tool [3, 7, 8].

1.1.3 Kohn–Sham Scheme

The Kohn–Sham (KS) scheme [2–4, 9] provides an avenue that the large part of the kinetic energy can be approached to good accuracy. This was, however, achieved by utilizing an orbital representation. It is assumed that, for any real (interacting) system, there exists a local single particle potential $v_s(\vec{r})$ corresponding to a fictitious (non-interacting fermion) system, whose ground state density $\rho(\vec{r})$ equals the exact ground state density $\rho_0(\vec{r})$. Such a system is defined by the Hamiltonian

$$H_s = \sum_{i=1}^{N} \left(-\frac{1}{2}\nabla_i^2 + v_s(\vec{r}_i) \right) \tag{1.14}$$

which has an exact wavefunction solution that is the single Slater determinant constructed from the N lowest orbitals of the one-electron equations

$$\left(-\frac{1}{2}\nabla^2 + v_s(\vec{r}) \right)\phi_i(\vec{r}) = \varepsilon_i\phi_i(\vec{r}) \tag{1.15}$$

For this system the kinetic energy and electron density are given by

$$T_s[\rho] = -\frac{1}{2}\sum_{i=1}^{N} \int \phi_i^*(\vec{r})\nabla^2\phi_i(\vec{r})\mathrm{d}\vec{r} \tag{1.16}$$

$$\rho(\vec{r}) = \sum_{i=1}^{N} |\phi_i(\vec{r})|^2 \tag{1.17}$$

and its total energy is given by

$$E_s[\rho] = T_s[\rho] + \int v_s(\vec{r})\rho(\vec{r})\mathrm{d}\vec{r} \qquad (1.18)$$

The corresponding Euler–Lagrange equation is thus

$$\mu = v_s(\vec{r}) + \frac{\delta T_s[\rho]}{\delta\rho(\vec{r})} \qquad (1.19)$$

The quantity $T_s[\rho]$, although not the exact kinetic energy $T[\rho]$, is well-defined using orbitals $\{\phi_i\}$. It is a density functional, because the KS orbitals are implicit functionals of density. T_s is the major contributor of $T[\rho]$. Its proper description is the key to the success of KS-DFT. Nevertheless, one has to notice that, by reintroducing the wavefunction, the full potential of DFT in having only three variables independent of system size could no longer be realized.

Now $F_{HK}[\rho]$ in Eq. 1.10 can be reformulated as

$$F_{HK}[\rho] = T_s[\rho] + J[\rho] + E_{xc}[\rho] \qquad (1.20)$$

and the corresponding Euler–Lagrange (Eq. 1.13) becomes

$$\mu = v_{ext}(\vec{r}) + \frac{\delta T_s[\rho]}{\delta\rho(\vec{r})} + \frac{\delta J[\rho]}{\delta\rho(\vec{r})} + \frac{\delta E_{xc}[\rho]}{\delta\rho(\vec{r})} \qquad (1.21)$$

Here $E_{xc}[\rho]$ is the so-called exchange-correlation (xc) functional, which covers the residual part of the true kinetic energy ($T_c = T - T_s$) and the non-classic electrostatic interaction of electrons ($E_{ncl} = V_{ee} - J$).

$$E_{xc}[\rho] = (T[\rho] - T_s[\rho]) + (V_{ee}[\rho] - J[\rho]) \qquad (1.22)$$

The exact xc functional $E_{xc}[\rho]$ is unknown, whose construction should be a relatively "easy" job as compared to the direct construction of $F_{HK}[\rho]$, as the former is a minor, while the latter is a major, contributor to the total energy functional $E[\rho]$.

Recall that the KS scheme starts by assuming the existence of $v_s(\vec{r})$, which remains to be determined. By requiring the chemical potential of the non-interacting system (Eq. 1.19) to be equal to that of the real system (Eq. 1.21), $v_s(\vec{r})$ can then be obtained as

$$v_s(\vec{r}) = v_{ext}(\vec{r}) + v_J(\vec{r}) + v_{xc}(\vec{r}) \qquad (1.23)$$

where the external potential is given in Eq. 1.4, and the Coulomb as well as the xc potentials are given by

$$v_J(\vec{r}) = \frac{\delta J[\rho]}{\delta\rho(\vec{r})} = \int \frac{\rho(\vec{r}')}{|\vec{r}' - \vec{r}|}\mathrm{d}\vec{r}' \qquad (1.24)$$

$$v_{xc}(\vec{r}) = \frac{\delta E_{xc}[\rho]}{\delta\rho(\vec{r})} \qquad (1.25)$$

Note that $v_s(\vec{r})$ depends on ρ, via Eqs. (1.23)–(1.25), the KS Equation (1.15) must be solved iteratively. Furthermore, the exact xc functional is unknown, such that the exact xc potential, defined in Eq. 1.25, is unknown. Certain approximated $E_{xc}[\rho]$ has to be used in practice. Thus, pursuing more and more accurate and reliable approximate xc functionals is the key issue in the development of density functional theory.

1.2 Approximations for the Commonly Used Exchange-Correlation Functionals

1.2.1 The Jacob's Ladder

From the above section, we see that the Kohn–Sham scheme allows us to separate out $E_{xc}[\rho]$ from $F_{HK}[\rho]$. This leaves finding an accurate form for $E_{xc}[\rho]$ the last barrier to the finding of the ground state density and the ground state energy.

Various density functional approximations (DFAs) to the xc energy have been developed in recent decades [9–56]. They are proposed with different philosophies. Functionals may be categorized into non-empirical and empirical. The former are formulated only by satisfying some physical rules, (e.g. [9–16, 21–23, 31, 32]) while the later are made by fitting to the known results of atomic or molecular properties (e.g. [25, 26, 33, 38, 39, 45–48]). In practice, the most popular functionals seem to be those developed by a combination of these two approaches, i.e., a physically motivated form with a few parameters being optimized for better numerical performance (e.g. [19, 20, 28, 29, 34, 35, 43, 50, 54]).

Perdew proposed a useful scheme to categorize the existing functionals, which also points out the direction for future functional development. This scheme is known as the Jacob's ladder (See Fig. 1.1), where functionals are grouped according to their complexity on rungs of the ladder, starting from the Hartree approximation on "earth" to the "heaven of chemical accuracy" [57, 58].

We will now briefly discuss the first four rungs of this ladder to introduce some of the most widely used xc functionals in the past and present. The focus of the present book is to introduce, from the next chapter, a new generation of functionals on the fifth rung.

1.2.2 The First Rung Functionals

Local density approximation (LDA [9–16]) is the foundation of most DFAs, representing the first rung of the Jacob's ladder. In this approach, the real inhomogeneous system is divided into infinitesimal volumes. In each of the volumes, the electron density is taken to be constant. The xc energy for the density within

Fig. 1.1 Jacob's ladder of approximate DFT methods

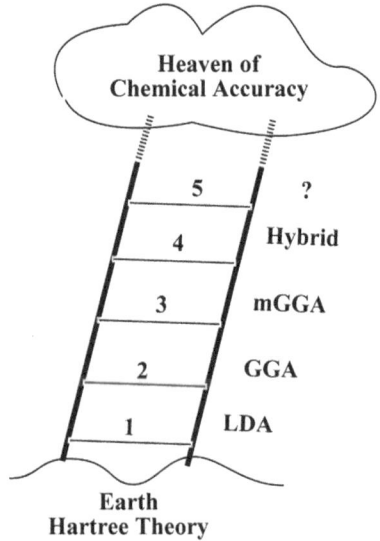

each volume is then assumed to be the xc energy obtained from the uniform electron gas for that density. Thus, the total xc energy of the system can be written as

$$E_{xc}^{LDA}[\rho] = \int \rho(\vec{r}) \varepsilon_{xc}^{LDA}([\rho]; \vec{r}) d\vec{r} \tag{1.26}$$

where ε_{xc}^{LDA} is the xc energy per particle.

It is usual to decompose $E_{xc}[\rho]$ into its exchange $E_x[\rho]$ and correlation $E_c[\rho]$ components:

$$E_{xc}[\rho] = E_x[\rho] + E_c[\rho] \tag{1.27}$$

For LDA, the analytical expression for the exchange energy is known by Bloch and Dirac [11, 12].

$$E_x^{LDA}[\rho] = -\frac{3}{4}\left(\frac{3}{\pi}\right)^{1/3} \int \rho(\vec{r})^{4/3} d\vec{r} \tag{1.28}$$

and

$$\varepsilon_x^{LDA} = -\frac{3}{4}\left(\frac{3}{\pi}\right)^{1/3} \rho(\vec{r})^{1/3} \tag{1.29}$$

Slater proposed the X_α method which was once widely used in the solid state physics [13]. Although it actually differs from Eq. 1.28 for the prefactor, the name Slater is often used as a synonym for the LDA exchange energy, E_x^S, involving the electron density raised to the 4/3 power.

In the literature, Eq. 1.29 is also often expressed in terms of the Wigner–Seitz radius r_s [59], which corresponds to the radius of the effective volume containing one electron:

$$r_s = \left(\frac{3}{4\pi\rho}\right)^{1/3} \tag{1.30}$$

Thus

$$\varepsilon_x^S = -\frac{3}{4}\left(\frac{3}{2\pi}\right)^{2/3}\frac{1}{r_s} \tag{1.31}$$

The correlation energy is more complicated. It has been derived in the high and low density limits [60–63]. For the intermediate densities, the correlation energy has been determined to a high precision by quantum Monte Carlo (QMC) methods [64]. In order to use these results in DFT calculations, it is desirable to have a suitable analytic interpolation formula. Several such formulas have been constructed and all are considered as accurate fits [14–16]. The common LDA correlation functionals are Vosko, Wilk, and Nusair (VWN) [14], and Perdew-Zunger (PL) [15], and Perdew–Wang [16].

Below, we will discuss more on the VWN parameterization, which is given in Eq. 1.32 [14].

$$\begin{aligned}
\varepsilon_c^{P/F} = A\Bigg[&\ln\frac{y^2}{Y(y)} + \frac{2b}{Q}\tan^{-1}\frac{Q}{2y+b} \\
&- \frac{by_0}{Y(y_0)}\left(\ln\frac{(y-y_0)^2}{Y(y)} + \frac{2(b+2y_0)}{Q}\tan^{-1}\frac{Q}{2y+b}\right)\Bigg]
\end{aligned} \tag{1.32}$$

where $y = r_s^{1/2}$, $Y(y) = y^2 + by + c$, $Q = (4c - b^2)^{1/2}$ and P/F stands for the para/ ferro magnetic case, respectively. For the paramagnetic case, the parameters fitted to QMC data are $A = 0.0310907$ (a.u.), $y_0 = -0.10498$, $b = 3.72744$, $c = 12.9352$; while for the ferromagnetic case, the corresponding parameters are $2A = 0.0310907$ (a.u.), $y_0 = -0.32500$, $b = 7.06042$, $c = 18.0578$. In the original VWN paper [14], parameters were also fitted to the Random Phase Approximation (RPA), where $y_0 = -0.409286$, $b = 13.0720$, $c = 42.7198$ for the paramagnetic case, and $y_0 = -0.743294$, $b = 20.1231$, $c = 101.578$ for the ferromagnetic case (with the same setting for A as in QMC). Note that different implementations have used different parameterizations which produce different numerical results and have caused some confusion. For example, Gaussian suite of program [65] uses the synonym VWN for the RPA parameterization and VWN5 for the QMC parameterization.

The extension of density functionals to spin-polarized systems is straightforward for exchange, where the exact spin-scaling is known as [66]:

$$E_x[\rho_\alpha, \rho_\beta] = \frac{1}{2}\left(E_x[2\rho_\alpha] + E_x[2\rho_\beta]\right) \tag{1.33}$$

and

$$\rho_\alpha + \rho_\beta = \rho \tag{1.34}$$

But for correlation further approximations must be employed. The spin-dependence of the correlation energy density, $\varepsilon_c(r_s, \zeta)$, is approached by introducing the relative spin-polarization [14]:

$$\zeta = \frac{\rho_\alpha(\vec{r}) - \rho_\beta(\vec{r})}{\rho(\vec{r})} \tag{1.35}$$

where $\zeta = 0$ corresponds to the paramagnetic spin-unpolarized situation $\varepsilon_c^P(r_s) = \varepsilon_c(r_s, \zeta = 0)$ with $\rho_\alpha = \rho_\beta$, whereas $\zeta = \pm 1$ corresponds to the ferromagnetic situation $\varepsilon_c^F(r_s) = \varepsilon_c(r_s, \zeta = \pm 1)$ where one spin density vanishes. The ζ-dependent correlation energy density $\varepsilon_c(r_s, \zeta)$ is then constructed to interpolate these extreme values. Again, different interpolation formulas may be employed. For example, the Gaussian implementation [65] of VWN uses the formula suggested by von Barth and Hedin [10], while its VWN5 implementation uses a slightly more complicated form where a quality called spin stiffness, $\alpha_c(r_s)$, was also fitted against Eq. 1.32 [14].

Thus we now have some commonly used acronyms for the LDA exchange-correlation functionals, namely SVWN, SVWN5, SPL as in Gaussian.

1.2.3 The Second Rung Functionals

LDA is best suited in describing extended systems, such as metals, where the approximation of the uniform electron gas is most valid. To account for the rapidly varying electron densities of many other systems like atoms and molecules, improvements over LDA have to make the xc energies dependent not only on the electron density but also on derivatives of the density.

The second rung functionals are based on the so-called generalized gradient approximation (GGA) [17–20], where the first derivative of the density is also included as a variable.

$$E_{xc}^{GGA}[\rho] = \int \rho(\vec{r})\varepsilon_{xc}^{GGA}([\rho, |\nabla\rho|]; \vec{r})\mathrm{d}\vec{r} \tag{1.36}$$

Similarly, a GGA xc functional is also assumed to be separable, and the GGA exchange functional takes the general form as

$$\varepsilon_x^{GGA}(\rho, |\nabla\rho|) = \varepsilon_x^{LDA} \cdot F(s) \tag{1.37}$$

where $F(s)$ is an enhancement factor and s (or similarly $x = (24\pi^2)^{\frac{1}{3}} \cdot s$) is the dimensionless reduced gradient defined as

$$s = \frac{|\nabla \rho|}{(24\pi^2)^{\frac{1}{3}} \rho^{\frac{4}{3}}} \tag{1.38}$$

There are now many exchange functionals of the GGA type (e.g. [17–29]). They differ from each other in the form of the enhancement factor $F(s)$. One of the most popular GGA exchange functionals was proposed by Becke, denoted as B or B88, whose $F(s)$ takes the form as [19]

$$F^{B88}(s) = \frac{1 + s \cdot a_1 \cdot \sinh^{-1}(s \cdot a_2) + a_3 \cdot s^2}{1 + s \cdot a_1 \cdot \sinh^{-1}(s \cdot a_2)} \tag{1.39}$$

Another popular exchange functional is called PW91 (Perdew–Wang 1991) [21].

$$F^{PW91}(s) = \frac{1 + s \cdot a_1 \cdot \sinh^{-1}(s \cdot a_2) + \left(a_3 + a_4 \cdot e^{-100s^2}\right) s^2}{1 + s \cdot a_1 \cdot \sinh^{-1}(s \cdot a_2) + a_5 \cdot s^d} \tag{1.40}$$

Here $A_x = -\frac{3}{4} \left(\frac{3}{\pi}\right)^{1/3}$, $a_2 = (48\pi^2)^{\frac{1}{3}}$, $a_1 = 6 \cdot a_2$, $a_3 = -\frac{a_2^2}{2^{1/3} A_x} \cdot \beta$, $a_4 = \frac{10}{81} - a_3$, $a_5 = \frac{-a_2^4 \times 10^{-6}}{2^{1/3} A_x}$, and $d = 4$. B88 contains one fitting parameter β. Becke obtained $\beta = 0.0042$ from fitting to Hartree–Fock (HF) exchange energies for the noble gas atoms [19]. PW91 adopted the B88 form by adding a_4 and a_5 to satisfy some physical constraints. PW91 is considered as non-empirical as it is an analytic fit to a numerical GGA obtained by real-space cutoff of the spurious long-range parts of the second-order gradient expansion of the exchange-correlation hole [21, 22].

PBE (Perdew-Burke-Ernzerhof) [23] can be considered as a refinement of PW91. In PBE, the enhancement factor of the exchange functional is given by

$$F^{PBE}(s) = 1 + \kappa - \frac{\kappa}{\left(1 + \frac{\mu}{\kappa} s^2\right)} \tag{1.41}$$

where $\kappa = 0.804$ is set to the maximum value allowed by the local Lieb-Oxford bound [23, 67] on E_{xc} and $\mu = 0.21951$ is set to recover the linear response of the uniform gas such that the effective gradient coefficient for exchange cancels that for correlation.

Note that the GGA functional form is usually too restrict to fulfill some important conditions simultaneously. For example, it is known that as r approaches infinity, $\rho(\vec{r})$ approaches $\exp(-\alpha \vec{r})$ so that [19, 68]

$$\lim_{r \to \infty} \varepsilon_x = -\frac{\rho(\vec{r})}{2r} \tag{1.42}$$

Levy and Perdew showed that some scaling properties can be satisfied if the asymptotic form of the functional for large s is s^{-a}, where $a \geq 1/2$ (Condition 2) [69]

$$\lim_{s \to \infty} F(s) = \frac{1}{s^a} \left(a \geq \frac{1}{2} \right) \text{ (Condition 2)} \tag{1.43}$$

while the global version of the Lieb-Oxford bound requires that [23, 67]

$$E_x \geq E_{xc} \geq -1.679 \int \rho(\vec{r})^{\frac{4}{3}} d\vec{r} \text{ (Condition 3)} \tag{1.44}$$

Figure 1.2 depicts the $F(s)$ functions. B88 fulfills Condition 1 by construction [19], but violates Conditions 2 and 3. PW91 does not obey Condition 1, but fulfills the other two conditions by adding a_4 and a_5 [22]. PBE meets only Condition 3. It was chosen to sacrifice Condition 2 to avoid the $F(s)$ turnover of PW91, as it was suspected that this turnover would cause spurious wiggles in the exchange potential for large s [23]. While all three functionals recover the LDA limit as $s \to 0$, the large s behavior cannot be uniquely determined (see Fig. 1.2). As the density is typically low and varies rapidly in the nonbonded region, proper large s behavior is important for the description of the nonbonded interactions [43].

There are similarly various GGA functionals being proposed for the correlation energy (e.g. [20, 22, 23]). The GGA correlation functional is usually expressed as:

$$E_c^{GGA}\left[\rho_\alpha, \rho_\beta\right] = \int \rho(\vec{r})\left[\varepsilon_c^{LDA}(r_s, \zeta) + H(r_s, \zeta, t)\right]d\vec{r} \tag{1.45}$$

where $t = \dfrac{|\nabla \rho|}{2gk_s\rho}$ is another scaled density gradient, $g = \frac{1}{2}\left[(1 + \zeta)^{\frac{2}{3}} + (1 - \zeta)^{\frac{2}{3}}\right]$ is a spin-scaling factor and $k_s = \left(\frac{4k_F}{\pi}\right)^{\frac{1}{2}}$ is the Thomas–Fermi screening wave vector with $k_F = (3\pi^2\rho)^{\frac{1}{3}}$ being the local Fermi wave vector.

Fig. 1.2 Enhancement factors for a set of GGA exchange functionals

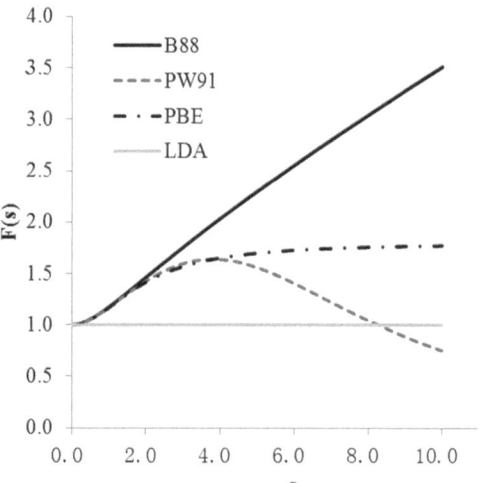

In the Perdew–Wang-91 correlation functional (E_c^{PW91}), H is expanded as [22]

$$H^{PW91} = H_0 + H_1 \tag{1.46}$$

$$H_0 = g^3 \frac{\beta^2}{2\alpha} \ln\left[1 + \frac{2\alpha}{\beta} \frac{t^2 + At^4}{1 + At^2 + A^2 t^4}\right] \tag{1.47}$$

$$H_1 = \left(\frac{16}{\pi}\right) (3\pi^2)^{\frac{1}{3}} \left[C_c(r_s) - C_c(0) - \frac{3C_x}{7}\right] g^3 t^2 \exp\left[-100 g^4 t^2 \left(\frac{k_s^2}{k_F^2}\right)\right] \tag{1.48}$$

with parameters $A = \dfrac{2\alpha}{\beta} \dfrac{1}{\exp\left[\dfrac{-2\alpha \varepsilon_c^{LDA}(r_s, \zeta)}{g^3 \beta^2}\right] - 1}$, $\alpha = 0.09$, $\beta = 0.066725$, the

Rasolt and Geldart constants (C_c [70]) and the Sham coefficient (C_x [71]). ε_c^{LDA} is the Perdew–Wang's parameterization ε_c^{PW} [16].

In the well-established PBE correlation functional, only the first term in the PW91 correlation functional is kept [21–23]. This was derived to ensure that the PBE correlation functional reduces to the correct second-order gradient expansion in the slowly varying limit, and under uniform scaling of the density $\left[\rho_\gamma(\vec{r}) = \gamma^3 \rho(\gamma \vec{r})\right]$ the PBE correlation energy correctly scales to a constant in the limit $\gamma \to \infty$. In addition PBE uses $\alpha = 0.0716$ instead of the $\alpha = 0.09$ used in PW91. Note that the PBE form has been revised [27, 28] and the parameters have been optimized [28, 29].

Another popular GGA correlation functional is called LYP (Lee–Yang–Parr) [20].

$$E_c^{LYP}[\rho_\alpha, \rho_\beta] = -a \int \frac{4}{1 + d\rho^{-1/3}} \frac{\rho_\alpha \rho_\beta}{\rho} d\vec{r}$$

$$-ab \int \omega d\vec{r} \left\{ \rho_\alpha \rho_\beta \left[2^{11/3} C_F \left(\rho_\alpha^{8/3} + \rho_\beta^{8/3}\right) + \left(\frac{47}{18} - \frac{7}{18}\delta\right) |\nabla\rho|^2 \right. \right.$$

$$- \left(\frac{5}{2} - \frac{1}{18}\delta\right)\left(|\nabla\rho_\alpha|^2 + |\nabla\rho_\beta|^2\right) - \frac{\delta - 11}{9} \left(\frac{\rho_\alpha}{\rho} |\nabla\rho_\alpha|^2 + \frac{\rho_\beta}{\rho} |\nabla\rho_\beta|^2\right) \right]$$

$$\left. - \frac{2}{3}\rho^2 |\nabla\rho|^2 + \left(\frac{2}{3}\rho^2 - \rho_\alpha^2\right)|\nabla\rho_\beta|^2 + \left(\frac{2}{3}\rho^2 - \rho_\beta^2\right)|\nabla\rho_\alpha|^2 \right\} \tag{1.49}$$

with parameters $\omega(\rho) = \dfrac{e^{-c\rho^{-1/3}}}{1 + d\rho^{-1/3}} \rho^{-11/3}$, $\delta(\rho) = c\rho^{-1/3} + \dfrac{d\rho^{-1/3}}{1 + d\rho^{-1/3}}$,

$a = 0.04918$, $b = 0.132$, $c = 0.2533$, $d = 0.349$, and $C_F = (3/10)(3\pi^2)^{2/3}$.

LYP does not use the electron gas as the reference system, but considers instead short-range effects in the two-particle density matrix. It is a simplification of the Colle-Salvetti correlation functional which explicitly depends on the single-particle orbitals [72]. LYP contains four semiempirical parameters $\{a, b, c, d\}$, all from the underlying Colle-Salvetti functional and obtained from a fit to the He atom. Note that LYP is self-interaction free for any one-electron system, but its correlation energy erroneously vanishes for any ferromagnetic situation.

The LYP correlation functional is often combined with the B88 exchange functional, while the PBE correlation functional is meant to be combined with its own exchange functional to ensure a good behavior of the whole xc functional. Thus we now have three commonly used acronyms for the GGA exchange-correlation functionals, namely BLYP, BPW91, and PBE.

1.2.4 The Third Rung Functionals

The third rung functionals are the so-called meta-GGAs [30–33], which logically extend the GGA functionals by allowing the xc functionals to depend further on Laplacian $\nabla^2\rho(r)$, the second order derivative of the electron density.

$$E_{xc}^{meta-GGA}[\rho] = \int \rho(\vec{r})\varepsilon_{xc}^{meta-GGA}\big([\rho, |\nabla\rho|, \nabla^2\rho, \tau]; \vec{r}\big)\mathrm{d}\vec{r} \qquad (1.50)$$

More frequently, meta-GGAs include τ as the additional ingredient, which is the orbital kinetic energy density defined by

$$\tau_\sigma(\vec{r}) = \frac{1}{2}\sum_i^{occ} |\nabla\phi_{i\sigma}(\vec{r})|^2 \qquad (1.51)$$

Here σ stands for either α or β spin and $\tau = \sum_\sigma \tau_\sigma$, while occ refers to occupied orbitals. The orbital kinetic energy density and the Laplacian of the density essentially carry the same information, since they are related via the orbitals and the effective potential of the KS equation [73, 74].

$$\tau_\sigma(\vec{r}) = \sum_i^{occ} \varepsilon_{i\sigma}|\phi_{i\sigma}(\vec{r})|^2 - v_s^\sigma(\vec{r})\rho_\sigma(\vec{r}) + \frac{1}{4}\nabla^2\rho_\sigma(\vec{r}) \qquad (1.52)$$

This may also be seen from the gradient expansion for slowly varying densities [31].

$$\tau = \tau^{TF} + \frac{1}{9}\tau^W + \frac{1}{6}\nabla^2\rho + O(\nabla^4\rho) \qquad (1.53)$$

with $\tau^{TF} = \frac{3}{10}(3\pi^2)^{2/3}\rho^{5/3}$ [2, 3, 13] the Thomas–Fermi kinetic energy density from uniform electron gas and $\tau^W = \dfrac{|\nabla\rho|^2}{8\rho}$ [31, 73–75] the von Weizsäcker kinetic energy density. As the representative meta-GGAs, we will introduce VSXC (van Voorhis-Scuseria) and TPSS (Tao-Perdew-Staroverov-Scuseria). While the former contains 21 parameters which are fitted to experimental data [30], the latter is a non-empirical version that improves over the PBE functional [32].

In VSXC, the exchange functional is expressed as [30]

$$E_x^{VSXC}[\rho_\alpha, \rho_\beta] = \sum_\sigma \int \rho_\sigma^{3/4} f(x_\sigma, z_\sigma) d\vec{r} \tag{1.54}$$

Here x_σ and z_σ are defined as $x_\sigma = \dfrac{|\nabla \rho_\sigma|}{\rho_\sigma^{4/3}}$ and $z_\sigma = 2\left(\dfrac{\tau_\sigma}{\rho_\sigma^{5/3}} - C_F\right)$, respectively.
This form was based on a density-matrix expansion, which was then modified through the introduction of 7 fit parameters $\{a, b, c, d, e, f, \alpha\}$ [30].

$$f(x_\sigma, z_\sigma) = \frac{a}{\gamma(x_\sigma, z_\sigma)} + \frac{bx_\sigma^2 + cz_\sigma}{\gamma^2(x_\sigma, z_\sigma)} + \frac{dx_\sigma^4 + ex_\sigma^2 z_\sigma + fz_\sigma^2}{\gamma^3(x_\sigma, z_\sigma)} \tag{1.55}$$

with $\gamma(x_\sigma, z_\sigma) \equiv 1 + \alpha(x_\sigma^2 + z_\sigma)$.
The correlation functional was constructed as

$$E_c^{VSXC} = E_c^{\alpha\beta} + E_c^{\alpha\alpha} + E_c^{\beta\beta} \tag{1.56}$$

$$E_c^{\sigma\sigma'} = \int \varepsilon_c^{LDA,\,\sigma\sigma'} f(x, z) d\vec{r} \tag{1.57}$$

$$E_c^{\sigma\sigma} = \int \varepsilon_c^{LDA,\,\sigma\sigma} D_\sigma f(x_\sigma, z_\sigma) d\vec{r} \tag{1.58}$$

where $x^2 \equiv x_\alpha^2 + x_\beta^2$ and $z \equiv z_\alpha + z_\beta$. ε_c^{LDA} is the Perdew–Wang's parameterization ε_c^{PW} with the spin components as defined by Stoll et al. [76]. The enhancement factor $f(x_\sigma, z_\sigma)$ takes the same form as in the exchange functional. The fit parameters are listed in Table 1.1 [30].

Note that D_σ is a dimensionless factor given by

$$D_\sigma = 1 - \frac{x_\sigma^2}{4(z_\sigma + 2C_F)} \tag{1.59}$$

It is designed to be zero for any system with only one orbital, which guarantees that the correlation energy is zero for any one-electron system to be self-interaction error free.

In TPSS, the exchange functional is defined as [32]

$$E_x^{TPSS}[\rho] = \int \rho(\vec{r}) \varepsilon_x^{LDA}([\rho]; \vec{r}) F^{TPSS}(p, \tilde{z}) d\vec{r} \tag{1.60}$$

The enhancement factor $F_x^{TPSS}(p, z)$ is given by

$$F^{TPSS}(p, \tilde{z}) = 1 + \kappa - \frac{\kappa}{\left(1 + \frac{\tilde{x}}{\kappa}\right)} \tag{1.61}$$

where μs^2 in F^{PBE} (Eq. 1.41) is replaced by a function $\tilde{x}$.

Table 1.1 Optimized fit parameters for VSXC [30]

	a	b	c	d	e	f	α
E_x	-9.800683×10^{-1}	-3.556788×10^{-3}	6.250326×10^{-3}	-2.354518×10^{-5}	-1.283732×10^{-4}	3.574822×10^{-4}	1.86726×10^{-3}
$E_c^{\alpha\alpha}$	3.270912×10^{-1}	-3.228915×10^{-2}	-2.942406×10^{-2}	2.134222×10^{-3}	5.451559×10^{-3}	1.577575×10^{-2}	5.15088×10^{-3}
$E_c^{\alpha\alpha'}$	7.035010×10^{-1}	7.694574×10^{-3}	5.152765×10^{-2}	3.394308×10^{-5}	-1.269420×10^{-3}	1.296118×10^{-3}	3.04966×10^{-3}

$$\tilde{x} = \left\{ \left[\frac{10}{81} + c \frac{\tilde{z}^2}{(1+\tilde{z}^2)^2} \right] p + \frac{146}{2025} \tilde{q}_b^2 - \frac{73}{405} \tilde{q}_b \sqrt{\frac{1}{2} \left(\frac{3}{5} \tilde{z} \right)^2 + \frac{1}{2} p^2} \right.$$
$$\left. + \frac{1}{\kappa} \left(\frac{10}{81} \right)^2 p^2 + 2\sqrt{e} \frac{10}{81} \left(\frac{3}{5} \tilde{z} \right)^2 + e\mu p^3 \right\} (1 + \sqrt{e}p)^{-2} \qquad (1.62)$$

with $\tilde{z} = \tau^W / \tau$, $\alpha = \dfrac{\tau - \tau^W}{\tau^{TF}}$, $p = s^2$ the square of the reduced gradient, and $\tilde{q}_b = \dfrac{9}{20} \dfrac{(\alpha-1)}{[1+b\cdot\alpha(\alpha-1)]^{1/2}} + \dfrac{2p}{3}$. Here $\mu = 0.21951$, $\kappa = 0.804$, $c = 1.59096$ and $e = 1.537$ were fixed from some physical arguments [32], while b was chosen to be 0.40 [71]. Note that we have used symbols $\tilde{x}$ and $\tilde{z}$ here, rather than symbols x and z used in the original paper [32], as the latter have already been used in defining the VSXC functional [30].

In TPSS, the correlation functional is given by [32]

$$E_c^{TPSS}[\rho] = \int \rho(\vec{r}) \varepsilon_c^{revPKZB} \left[1 + d\tilde{z}^3 \varepsilon_c^{revPKZB} \right] d\vec{r} \qquad (1.63)$$

with

$$\varepsilon_c^{revPKZB} = \varepsilon_c^{PBE} \left[1 + C(\zeta,\xi)\tilde{z}^2 \right] - [1 + C(\zeta,\xi)]\tilde{z}^2 \sum_\sigma \frac{\rho_\sigma}{\rho} \tilde{\varepsilon}_c \qquad (1.64)$$

$$C(\zeta,\xi) = \frac{C(0,0) + 0.87\zeta^2 + 0.50\zeta^4 + 2.26\zeta^6}{\left\{ 1 + \frac{\xi^2}{2} \left[\frac{1}{(1+\zeta)^{4/3}} + \frac{1}{(1-\zeta)^{4/3}} \right] \right\}^4} \qquad (1.65)$$

$$\tilde{\varepsilon}_c^\sigma = \max\{ \varepsilon_c^{PBE}(\rho_\sigma, 0, \nabla\rho_\sigma, 0), \varepsilon_c^{PBE}(\rho_\alpha, \rho_\beta, \nabla\rho_\alpha, \nabla\rho_\beta) \} \qquad (1.66)$$

where ζ is the relative spin polarization (Eq. 1.35), $\xi = \dfrac{|\nabla\zeta|}{2(3\pi^2\rho)^{1/3}}$. $C(0,0)$ and d were chosen to be 0.53 and 2.8 au^{-1}, respectively [77]. Note that some parameters in TPSS has been optimized or revised [78].

1.2.5 The Fourth Rung Functionals

As in Eqs. 1.26, 1.36, and 1.50, the xc functional is defined as an integral over $\vec{r}$ of a function of $\vec{r}$ which we call an energy density. An LDA corresponds to a 'local functional of the density' where the energy density at $\vec{r}$ is determined only by the electron density at $\vec{r}$. An GGA is a 'semilocal functional of the density' in the sense that the energy density at $\vec{r}$ is determined by the electron density in an infinitesimal neighborhood of $\vec{r}$. As energy density at $\vec{r}$ which employs kinetic energy density τ is computed from the density and the orbitals in an infinitesimal neighborhood of $\vec{r}$, an meta-GGAs is also 'semilocal functionals of the density and the orbitals.' Such (semi)locality gives rise to much of the computational convenience of DFT. However, these functionals are best suited for the description of slow varying density. For a better description of a finite system, there is a demand for certain nonlocal component.

With the KS orbitals at hand, the exchange energy can be explicitly written as

$$E_x^{HF} = -\frac{1}{2}\sum_{ij}^{occ}\iint\frac{\phi_j^*(\vec{r}')\phi_i^*(\vec{r})\phi_j(\vec{r})\phi_i(\vec{r}')}{|\vec{r}'-\vec{r}|}d\vec{r}'d\vec{r} \quad (1.67)$$

The superscript HF recognizes the fact that it shares the same form as in the Hartree–Fock (HF) theory, which however uses the HF orbitals. E_x^{HF} is nonlocal which depends on two spatial variables (r, r').

It would be reasonable to expect that good results were obtainable if the approximate (semi)local exchange functionals were replaced with this exact equation for exchange energy:

$$E_{xc}^{DFA} = E_x^{HF} + E_c^{DFA} \quad (1.68)$$

Regretfully, this form only gave acceptable results for one-center systems (atoms and ions) [4, 34, 35, 79], and was not very successful in describing chemical bonds in molecules [4, 34, 35, 80]. It is difficult to develop a (semi)local E_c^{DFA} that can match well with the nonlocal E_x^{HF}.

It should be emphasized that $J[\rho]$ as in Eq. 1.9 includes electron self-interaction explicitly. An approximated E_{xc}^{DFA} is unable to remove this self-interaction error completely, which has been related to many deficiencies of common DFAs [81–85]. This is the major advantage to use E_x^{HF}, which is one-electron self-interaction error free. On the other hand, as opposed to E_x^{HF}, the local or semilocal exchange functional E_x^{DFA} was shown to incorporate an mimic of 'static correlation' $\left(E_x^{DFA} = E_x^{HF} + E_c^{Sta}\right)$, while the corresponding approximation for correlation energy models 'dynamic correlation' $\left(E_c^{DFA} \approx E_c^{Dyn}\right)$ [26, 86].

$$E_{xc}^{DFA} = E_x^{DFA} + E_c^{DFA} \approx E_x^{HF} + E_c^{Sta} + E_c^{Dyn} \quad (1.69)$$

Hence, a correct choice that compromises the needs between elimination of self-interaction and inclusion of nondynamic correlation is to hybrid the right-hand sides of Eqs. 1.68 and 1.69. The simplest choice is a linear combination, giving rise to a new class of xc functionals denoted as the hybrid xc functionals [34, 35, 37].

$$E_{xc}^{hyb} = a_0 E_x^{HF} + (1-a_0)E_x^{DFA} + E_c^{DFA} \quad (1.70)$$

This was first introduced by Becke in 1993 [34], who used a different theoretical rationale, namely the adiabatic connection path (see Sects. 2.2 and 2.4) [34, 35, 37, 50, 87–89]. Ideally, it would be desirable to optimize a_0 for each system and property, but Perdew, based on the accuracy of the fourth order perturbation theory for most molecules, suggests $a_0 = 0.25$ as the best single choice.

This has led to the PBE0 hybrid functional where E_{xc}^{DFA} is the PBE GGA functional [37, 40, 41].

One of the most popular functionals is based on Becke's three-parameter scheme (B3) [35].

$$E_{xc}^{B3hyb} = a_0 E_x^{HF} + (1 - a_0)E_x^{LDA} + a_x \Delta E_x^{GGA} + E_c^{LDA} + a_c \Delta E_c^{GGA} \qquad (1.71)$$

The parameters $\{a_0 = 0.20, \ a_x = 0.78, \ a_c = 0.81\}$ were obtained by fitting to 56 atomization energies, 42 ionization energies, 8 proton affinities, and 10 first row atomic energies. During his fit, Becke has originally chosen $E_{xc}^{GGA} = E_{xc}^{BPW91}$ [35]. This hybrid functional is then called B3PW91. Later it appeared [36] that a combination of the B3 scheme with the BLYP xc functional [19, 20] gave better results. This has brought about the most widely used functional in chemistry B3LYP. Although LYP was not separable as PW91 (i.e., Eq. 1.45 vs 1.49) by construction, it was assumed $\Delta E_c^{LYP} = E_c^{LYP} - E_c^{LDA}$ [90], where E_c^{LDA} can be the VWN5 parameterization in some implementations, or the VWN RPA parameterization in some other implementations such as in Gaussian [65]. This is often a source of confusion.

The success, as well as the failure, of B3LYP has initiated the development of many new hybrid functionals (e.g. [38–56]). A recent important development in DFT is the M06 family of functionals (M06, M06-2X and M06-HF, and M06-L) [33, 48], which currently provides the highest accuracy with a broad applicability for chemistry. The exchange for the M06 family consists of a linear combination of two terms: one term comes from the PBE exchange functional [23] multiplied by a kinetic-energy–density enhancement factor [33], and the other term is the VSXC exchange functional [30]. The correlation for the M06 family also involves two terms: one term is similar to the correlation functional in BMK [45] modified by Becke's self-interaction correction factor [91], the other term comes from the VSXC correlation functional [30]. M06, M06-2X, M06-HF are hybrid methods with increasing amount of HF exchange (0.27, 0.54, 1.00, respectively) while M06-L is a pure DFT, with around 40 parameters fitted against different data sets to emphasize different applications. Thus M06 is of general purpose, M06-2X is recommended for chemistry of the main group elements, M06-HF is for charge-transfer states in electronic spectroscopy, and M06-L for transition metal chemistry [33, 48].

1.2.6 General Trends for the Functional Performances Along the Jacob's Ladder

Table 1.2 gives a first glance for the performance of some traditional DFT methods as compared with some traditional wavefunction methods [44, 90]. These methods are still widely used ever since their constructions more than 20 years ago [92]. More complete assessment will be provided in Chap. 3.

Table 1.2 Performance for some properties obtained by some traditional DFT methods. Results of some wavefunction methods are also listed for comparison

	G2-1[a]			He$_2$	(H$_2$O)$_2$
	BDE(R$_e$)[b]	Dipole[c]	Harm. Freq[d]	ΔE(R$_e$)[e]	D$_e$(R$_{O...O}$)[f]
Wavefunction methods					
HF	82.0(0.022)	0.29	144	Unbound	1.73(−0.100)
MP2	23.7(0.014)	0.28	99	0.009(−0.02)	0.35(0.045)[g]
CCSD(T)	11.5(0.005)	0.10	31	0.002(−0.03)	0.42(0.036)[h]
DFT methods					
SVWN	43.5(0.017)	0.25	75	−0.229(0.593)	−3.58(0.238)
BPW91	6.0(0.014)	0.11	69	Unbound	1.84(0.002)
BLYP	9.6(0.013)	0.25	73	Unbound	1.27(−0.004)
B3PW91	4.8(0.008)	0.08	45	Unbound	1.40(0.025)
B3LYP	3.3(0.004)	0.09	32	Unbound	0.87(0.022)

[a] Mean absolute deviations (MADs) obtained by different methods [44, 90, 93, 94] for the 32 molecules belonging to the reduced G2 set [95]. [b] MADs for bond dissociation energies (BDEs) and bond distances in kcal/mol and Å, respectively. [c] MADs for dipole moments in Debye. [d] MADs for harmonic frequencies in cm^{-1}. [e] Errors (Ref.–Calc.) for binding energy and equilibrium distance in kcal/mol and Å, respectively. MP2 and CCSD(T) values at aug-cc-pV5Z are taken from Ref. [96]. Others are taken from Ref. [44]. The reference data, 0.022 kcal/mol and 2.970 Å, are taken from Ref. [97]. [f] Errors (Ref.–Calc.) for binding energy and equilibrium distance in kcal/mol and Å, respectively. The reference data, 5.44 kcal/mol and 2.948 Å, are taken from Ref. [44, 98, 99]. [g] From Ref. [100]. [h] From Ref. [101]

As shown by the data in Table 1.2, the LDA (SVWN) method gives good molecular structures and vibrational frequencies in the strongly bound systems as in the G2-1 set. The mean absolute deviations (MADs) associated with SVWN are 0.017 Å and 75 cm^{-1}, respectively, which is a significant improvement over HF (0.022 Å and 144 cm^{-1}) and is comparable to those of MP2 (0.014 Å and 99 cm^{-1}), while the latter is much more time-consuming and computational resource demanding. LDA does not work for hydrogen bonded (HB) systems as well as for van der Waals interactions. Besides, its errors for bond dissociation energies (BDEs) of covalent molecules (43.5 kcal/mol for the G2-1 set) are too big to be useful.

GGAs, as exemplified by data of BPW91 and BLYP in Table 1.2, improve over LDA dramatically on BDE calculations. MADs for the G2-1 set are of order 6-10 kcal/mol. They give increasingly satisfied molecular structures, especially for the HB systems, with MADs around 0.003 Å. However, GGAs apparently fail for van der Waals interactions.

The general trends for the improvement along LDA $\rightarrow$ GGA $\rightarrow$ Hybrid GGA are clearly seen in the description of covalently bound molecules of main group elements. For the G2-1 set, B3LYP only possesses an MAD of 0.004 Å for geometry predictions, being comparable to those obtained by the CCSD(T) method using basis set at the triple-zeta quality. B3LYP's error for BDEs is halved as compared to those of GGAs. However, along with GGAs, hybrid GGAs do not work for van der Waals interactions. All these functionals leave much room for further improvement.

References

1. McWeeny R (1992) Methods of molecular quantum mechanics. Academic Press, London
2. Perdew JP, Kurth S (2003) In: Fiolhais C, Nogueira F, Marques M (eds) A primer in density functional theory. Springer, Berlin
3. Parr RG, Yang WT (1989) Density functional theory of atoms and molecules. Oxford University Press, New York
4. Koch W, Holthausen MC (2001) A chemist's guide to density functional theory, 2nd edn. Wiley-VCH, New York
5. Hohenberg P, Kohn W (1964) Inhomogeneous electron gas. Phys Rev B 136:B864–B871. doi:10.1103/PhysRev.136.B864
6. Levy M (1979) Universal variational functionals of electron densities, 1st-order density matrices, and natural spin-orbitals and solution of the V-representability problem. Proc Natl Acad Sci USA 76:6062–6065. doi:10.1073/pnas.76.12.6062
7. Teller E (1962) On stability of molecules in Thomas-Fermi theory. Rev Mod Phys 34:627–631. doi:10.1103/RevModPhys.34.627
8. Lieb E, Simon B (1977) Thomas-Fermi theory of atoms, molecules and solids. Adv Math 23:22–116. doi:10.1016/0001-8708(77)90108-6
9. Kohn W, Sham LJ (1965) Self-consistent equations including exchange and correlation effects. Phys Rev 140:1133–1138. doi:10.1103/PhysRev.140.A1133
10. von Barth U, Hedin L (1972) A local exchange-correlation potential for the spin polarized case i. J Phys C: Solid State Phys 5:1629–1642. doi:10.1088/0022-3719/5/13/012
11. Bloch F (1929) Note to the electron theory of ferromagnetism and electrical conductivity. Z Phys 57:545–555. doi:10.1007/BF01340281
12. Dirac PAM (1930) Note on exchange phenomena in the Thomas atom. Math Proc Camb Phil Soc 26:376–385. doi:10.1017/S0305004100016108
13. Slater JC (1960) Quantum theory of atomic structure, vol 2. McGraw-Hill, New York
14. Vosko SH, Wilk L, Nusair M (1980) Accurate spin-dependent electron liquid correlation energies for local spin-density calculations–a critical analysis. Can J Phys 58:1200–1211. doi:10.1139/p80-159
15. Perdew JP, Zunger A (1981) Self-interaction correction to density-functional approximations for many electron systems. Phys Rev B 23:5048–5079. doi:10.1103/PhysRevB.23.5048
16. Perdew JP, Wang Y (1992) Accurate and simple analytic representation of the electron-gas correlation energy. Phys Rev B 45:13244–13249. doi:10.1103/PhysRevB.45.13244
17. Langreth DC, Mehl M (1983) Beyond the local-density approximation in calculations of ground-state electronic-properties. Phys Rev B 28:1809–1834. doi:10.1103/PhysRevB.28
18. Perdew JP (1986) Density-functional approximation for the correlation-energy of the inhomogeneous electron-gas. Phys Rev B 33:8822–8824. doi:10.1103/PhysRevB.33.8822; ibid. (1986) 34:7406 (E)
19. Becke AD (1988) Density-functional exchange-energy approximation with correct asymptotic behavior. Phys Rev A 38:3098–3100. doi:10.1103/PhysRevA.38.3098
20. Lee CT, Yang WT, Parr RG (1988) Development of the Colle-Salvetti correlation-energy formula into a functional of the electron-density. Phys Rev B 37:785–789. doi:10.1103/PhysRevB.37.785
21. Perdew JP (1991) Electronic structure of solids'91. Akademie Verlag, Berlin
22. Perdew JP, Chevary JA, Vosko SH et al (1992) Atoms, molecules, solids, and surfaces: Applications of the generalized gradient approximation for exchange and correlation. Phys Rev B 46:6671–6687. doi:10.1103/PhysRevB.46.6671
23. Perdew JP, Burke K, Ernzerhof M (1996) Generalized gradient approximation made simple. Phys Rev Lett 77:3865–3868. doi:10.1103/PhysRevLett.77.3865

24. Adamo C, Barone V (1998) Exchange functionals with improved long-range behavior and adiabatic connection methods without adjustable parameters: The mPW and mPW1PW models. J Chem Phys 108:664–675. doi:10.1063/1.475428
25. Hamprecht FA, Cohen AJ, Tozer DJ, Handy NC (1998) Development and assessment of new exchange-correlation functionals. J Chem Phys 109:6264–6271. doi:10.1063/1.477267
26. Cohen AJ, Handy NC (2001) Dynamic correlation. Mol Phys 99:607–615. doi:10.1080/00268970010023435
27. Hammer B, Hansen LB, Nørskov JK (1999) Improved adsorption energetics within density-functional theory using revised Perdew-Burke-Ernzerhof functionals. Phys Rev B 59:7413–7421. doi:10.1103/PhysRevB.59.7413
28. Zhang YK, Yang WT (1998) Comment on "Generalized gradient approximation made simple". Phys Rev Lett 80:890–890. doi:10.1103/PhysRevLett.80.890
29. Xu X, Goddard WA (2004) The extended Perdew-Burke-Ernzerhof functional with improved accuracy for thermodynamic and electronic properties of molecular systems. J Chem Phys 121:4068–4082. doi:10.1063/1.1771632
30. Van Voorhis T, Scuseria GE (1998) A novel form for the exchange-correlation energy functional. J Chem Phys 109:400–410. doi:10.1063/1.476577
31. Perdew JP, Kurth S, Zupan A, Blaha P (1999) Accurate density functional with correct formal properties: A step beyond the generalized gradient approximation. Phys Rev Lett 82:2544–2547. doi:10.1103/PhysRevLett.82.2544
32. Tao JM, Perdew JP, Staroverov VN, Scuseria GE (2003) Climbing the density functional ladder: Nonempirical meta-generalized gradient approximation designed for molecules and solids. Phys Rev Lett 91:146401–146404. doi:10.1103/PhysRevLett.91.146401
33. Zhao Y, Truhlar DG (2006) A new local density functional for main-group thermochemistry, transition metal bonding, thermochemical kinetics, and noncovalent interactions. J Chem Phys 125:194101. doi:10.1063/1.2370993
34. Becke AD (1993) A new mixing of Hartree–Fock and local density-functional theories. J Chem Phys 98:1372–1377. doi:10.1063/1.464304
35. Becke AD (1993) Density-functional thermochemistry 3: The role of exact exchange. J Chem Phys 98:5648–5652. doi:10.1063/1.464913
36. Stephens PJ, Devlin FJ, Chabalowski CF, Frisch MJ (1994) Ab-initio calculation of vibrational absorption and circular-dichroism spectra using density-functional force-fields. J Phys Chem 98:11623–11627. doi:10.1021/j100096a001
37. Perdew JP, Emzerhof M, Burke K (1996) Rationale for mixing exact exchange with density functional approximations. J Chem Phys 105:9982–9985. doi:10.1063/1.472933
38. Becke AD (1997) Density-functional thermochemistry. 5. Systematic optimization of exchange-correlation functionals. J Chem Phys 107:8554–8560. doi:10.1063/1.475007
39. Schmider HL, Becke AD (1998) Optimized density functionals from the extended G2 test set. J Chem Phys 108:9624–9631. doi:10.1063/1.476438
40. Ernzerhof M, Scuseria GE (1999) Assessment of the Perdew-Burke-Ernzerhof exchange-correlation functional. J Chem Phys 110:5029–5036. doi:10.1063/1.478401
41. Adamo C, Barone V (1999) Toward reliable density functional methods without adjustable parameters: the PBE0 model. J Chem Phys 110:6158–6170. doi:10.1063/1.478522
42. Xu X, Goddard WA (2004) Assessment of Handy-Cohen optimized exchange density functional (OPTX). J Phys Chem A 108:8495–8504. doi:10.1021/jp047428v
43. Xu X, Goddard WA (2004) The X3LYP extended density functional for accurate descriptions of nonbond interactions, spin states, and thermochemical properties. Proc Natl Acad Sci USA 101:2673–2677. doi:10.1073/pnas.0308730100
44. Xu X, Zhang QS, Muller RP, Goddard WA (2005) An extended hybrid density functional (X3LYP) with improved descriptions of nonbond interactions and thermodynamic properties of molecular systems. J Chem Phys 122:014105. doi:10.1063/1.1812257
45. Boese AD, Martin JML (2004) Development of density functionals for thermochemical kinetics. J Chem Phys 121:3405–3416. doi:10.1063/1.1774975

46. Zhao Y, Lynch BJ, Truhlar DG (2004) Doubly hybrid meta DFT: New multi-coefficient correlation and density functional methods for thermochemistry and thermochemical kinetics. J Phys Chem A 108:4786–4791. doi:10.1021/jp049253v

47. Zhao Y, Truhlar DG (2005) Design of density functionals that are broadly accurate for thermochemistry, thermochemical kinetics, and nonbonded interactions. J Phys Chem A 109:5656–5667. doi:10.1021/jp050536c

48. Zhao Y, Truhlar DG (2008) The M06 suite of density functionals for main group thermochemistry, thermochemical kinetics, noncovalent interactions, excited states, and transition elements: two new functionals and systematic testing of four M06-class functionals and 12 other functionals. Theor Chem Acc 120:215–241. doi:10.1007/s00214-007-0310-x

49. Zhang Y, Wu AA, Xu X, Yan YJ (2006) OPBE: A promising density functional for the calculation of nuclear shielding constants. Chem Phys Lett 421:383–388. doi:10.1016/j.cplett.2006.01.095

50. Mori-Sánchez P, Cohen AJ, Yang WT (2006) Self-interaction-free exchange-correlation functional for thermochemistry and kinetics. J Chem Phys 124:091102. doi:10.1063/1.2179072

51. Grimme S (2006) Semiempirical hybrid density functional with perturbative second-order correlation. J Chem Phys 124:034108. doi:10.1063/1.2148954

52. Karton A, Tarnopolsky A, Lamere JF et al (2008) Highly accurate first-principles benchmark data sets for the parametrization and validation of density functional and other approximate methods. Derivation of a robust, generally applicable, double-hybrid functional for thermochemistry and thermochemical kinetics. J Phys Chem A 112:12868–12886. doi:10.1021/jp801805p

53. Chai J-D, Head-Gordon M (2009) Long-range corrected double-hybrid density functionals. J Chem Phys 131:174105. doi:10.1063/1.3244209

54. Zhang Y, Xu X, Goddard WA (2009) Doubly hybrid density functional for accurate descriptions of nonbond interactions, thermochemistry, and thermochemical kinetics. Proc Natl Acad Sci USA 106:4963–4968. doi:10.1073/pnas.0901093106

55. Zhang IY, Xu X, Jung Y, Goddard WA (2011) A fast doubly hybrid density functional method close to chemical accuracy using a local opposite spin ansatz. Proc Natl Acad Sci USA 108:19896–19900. doi:10.1073/pnas.1115123108

56. Goerigk L, Grimme S (2011) Efficient and accurate double-hybrid-meta-GGA density functionals—Evaluation with the extended GMTKN30 database for general main group thermochemistry, kinetics, and noncovalent interactions. J Chem Theory Comput 7:291–309. doi:10.1021/ct100466k

57. Perdew JP, Ruzsinszky A, Tao JM et al (2005) Prescription for the design and selection of density functional approximations: more constraint satisfaction with fewer fits. J Chem Phys 123:062201. doi:10.1063/1.1904565

58. Furche F, Perdew JP (2006) The performance of semilocal and hybrid density functionals in 3d transition-metal chemistry. J Chem Phys 124:044103. doi:10.1063/1.2162161

59. Wigner E, Seitz F (1934) On the constitution of metallic sodium II. Phys Rev 46:509–524. doi:10.1103/PhysRev.46.509

60. Gell-Mann M, Brueckner KA (1957) Correlation energy of an electron gas at high density. Phys Rev 106:364–368. doi:10.1103/PhysRev.106.364

61. Carr WJ, Maradudin AA (1964) Ground-state energy of a high-density electron gas. Phys Rev 133:A371–A374. doi:10.1103/PhysRev.133.A371

62. Nozières P, Pines D (1958) Correlation energy of a free electron gas. Phys Rev 111:442–454. doi:10.1103/PhysRev.111.442

63. Carr WJ (1961) Energy, specific heat, and magnetic properties of the low-density electron gas. Phys Rev 122:1437–1446. doi:10.1103/PhysRev.122.1437

64. Ceperley DM, Alder BJ (1980) Ground state of the electron gas by a stochastic method. Phys Rev Lett 45:566–569. doi:10.1103/PhysRevLett.45.566

65. Frisch MJ et al. (2003) Gaussian 03, revision A. 1. Gaussian, Inc, Pittsburgh

66. Oliver GL, Perdew JP (1979) Spin-density gradient expansion for the kinetic energy. Phys Rev A 20:397–403. doi:10.1103/PhysRevA.20.397
67. Lieb EH, Oxford S (1981) Improved lower bound on the indirect Coulomb energy. Int J Quantum Chem 19:427–439. doi:10.1002/qua.560190306
68. Della Sala F, Görling A (2002) Asymptotic behavior of the Kohn-Sham exchange potential. Phys Rev Lett 89:033003. doi:10.1103/PhysRevLett.89.033003
69. Levy M, Perdew JP (1993) Tight bound and convexity constraint on the exchange-correlation-energy functional in the low-density limit, and other formal tests of generalized-gradient approximations. Phys Rev B 48:11638–11645. doi:10.1103/PhysRevB.48.11638
70. Rasolt M, Geldart DJW (1986) Exchange and correlation energy in a nonuniform fermion fluid. Phys Rev B 34:1325–1328. doi:10.1103/PhysRevB.34.1325
71. Sham LJ (1971) Computational Methods in Band Theory. Plenum, New York
72. Colle R, Salvetti O (1975) Approximate calculation of the correlation energy for the closed shells. Theoret Chim Acta 37:329–334. doi:10.1007/BF01028401
73. Becke AD (1983) Hartree–Fock exchange energy of an inhomogeneous electron gas. Int J Quantum Chem 23:1915–1922. doi:10.1002/qua.560230605
74. Becke AD (1998) A new inhomogeneity parameter in density-functional theory. J Chem Phys 109:2092–2098. doi:10.1063/1.476722
75. Weizsäcker CF v (1935) Zur theorie der kernmassen. Z Physik 96:431–458. doi:10.1007/BF01337700
76. Stoll H, Pavlidou CME, Preuß H (1978) On the calculation of correlation energies in the spin-density functional formalism. Theoret Chim Acta 49:143–149. doi:10.1007/BF02399063
77. Svendsen PS, von Barth U (1996) Gradient expansion of the exchange energy from second-order density response theory. Phys Rev B 54:17402–17413. doi:10.1103/PhysRevB.54.17402
78. Perdew JP, Ruzsinszky A, Csonka GI et al (2009) Workhorse semilocal density functional for condensed matter physics and quantum chemistry. Phys Rev Lett 103:026403. doi:10.1103/PhysRevLett.103.026403
79. Lagowski JB, Vosko SH (1988) An analysis of local and gradient-corrected correlation energy functionals using electron removal energies. J Phys B: At Mol Opt Phys 21:203. doi:10.1088/0953-4075/21/1/016
80. Clementi E, Chakravorty SJ (1990) A comparative study of density functional models to estimate molecular atomization energies. J Chem Phys 93:2591–2602. doi:10.1063/1.458899
81. Cohen AJ, Mori-Sánchez P, Yang WT (2011) Challenges for density functional theory. Chem Rev 112:289–320. doi:10.1021/cr200107z
82. Merkle R, Savin A, Preuss H (1992) Singly ionized first–row dimers and hydrides calculated with the fully-numerical density-functional program numol. J Chem Phys 97:9216–9221. doi:10.1063/1.463297
83. Zhang YK, Yang WT (1998) A challenge for density functionals: Self-interaction error increases for systems with a noninteger number of electrons. J Chem Phys 109:2604–2608. doi:10.1063/1.476859
84. Gräfenstein J, Kraka E, Cremer D (2004) The impact of the self-interaction error on the density functional theory description of dissociating radical cations: Ionic and covalent dissociation limits. J Chem Phys 120:524–539. doi:10.1063/1.1630017
85. Ciofini I, Adamo C, Chermette H (2005) Self-interaction error in density functional theory: a mean-field correction for molecules and large systems. Chem Phys 309:67–76. doi:10.1016/j.chemphys.2004.05.034
86. Gritsenko OV, Schipper PRT, Baerends EJ (1997) Exchange and correlation energy in density functional theory: Comparison of accurate density functional theory quantities with traditional Hartree–Fock based ones and generalized gradient approximations for the molecules Li_2, N_2, F_2. J Chem Phys 107:5007–5015. doi:10.1063/1.474864

87. Levy M, March NH, Handy NC (1996) On the adiabatic connection method, and scaling of electron–electron interactions in the Thomas–Fermi limit. J Chem Phys 104:1989–1992. doi:10.1063/1.470954

88. Gunnarsson O, Lundqvist BI (1976) Exchange and correlation in atoms, molecules, and solids by the spin-density-functional formalism. Phys Rev B 13:4274–4298. doi:10.1103/PhysRevB.13.4274

89. Langreth DC, Perdew JP (1977) Exchange-correlation energy of a metallic surface: Wave-vector analysis. Phys Rev B 15:2884–2901. doi:10.1103/PhysRevB.15.2884

90. Adamo C, Barone V (1997) Toward reliable adiabatic connection models free from adjustable parameters. Chem Phys Lett 274:242–250. doi:10.1016/S0009-2614(97)00651-9

91. Becke AD (1996) Density-functional thermochemistry. 4. A new dynamical correlation functional and implications for exact-exchange mixing. J Chem Phys 104:1040–1046. doi:10.1063/1.470829

92. Sousa SF, Fernandes PA, Ramos MJ (2007) General performance of density functionals. J Phys Chem A 111:10439–10452. doi:10.1021/jp0734474

93. Johnson BG, Gill PMW, Pople JA (1993) The performance of a family of density functional methods. J Chem Phys 98:5612–5626. doi:10.1063/1.464906

94. Johnson BG, Gonzales CA, Gill PMW, Pople JA (1994) A density functional study of the simplest hydrogen abstraction reaction. Effect of self-interaction correction. Chem Phys Lett 221:100–108. doi:10.1016/0009-2614(94)87024-1

95. Curtiss LA, Raghavachari K, Trucks GW, Pople JA (1991) Gaussian-2 theory for molecular-energies of 1st-row and 2nd-row compounds. J Chem Phys 94:7221–7230. doi:10.1063/1.460205

96. Roy D, Marianski M, Maitra NT, Dannenberg JJ (2012) Comparison of some dispersion-corrected and traditional functionals with CCSD(T) and MP2 ab initio methods: Dispersion, induction, and basis set superposition error. J Chem Phys 137:134109. doi:10.1063/1.4755990

97. Ogilvie JF, Wang FYH (1992) Potential-energy functions of diatomic molecules of the noble gases I. Like nuclear species. J Mol Struct 273:277–290. doi:10.1016/0022-2860(92)87094-C

98. Odutola JA, Dyke TR (1980) Partially deuterated water dimers: microwave spectra and structure. J Chem Phys 72:5062–5070. doi:10.1063/1.439795

99. Curtiss LA, Frurip DJ, Blander M (1979) Studies of molecular association in H_2O and D_2O vapors by measurement of thermal conductivity. J Chem Phys 71:2703–2711. doi:10.1063/1.438628

100. Taketsugu T, Wales DJ (2002) Theoretical study of rearrangements in water dimer and trimer. Mol Phys 100:2793–2806. doi:10.1080/00268970210142648

101. Klopper W, Rijdt JGCM van D de, Duijneveldt FB van (2000) Computational determination of equilibrium geometry and dissociation energy of the water dimer. Phys Chem Chem Phys 2:2227–2234. doi:10.1039/A910312K

Chapter 2
A New Generation of Doubly Hybrid Density Functionals (DHDFs)

Abstract Doubly hybrid density functionals (DHDFs) present a new generation of density functionals, which not only enfold a nonlocal orbital-dependent component (i.e., the Hartree-Fock-like exchange) in the exchange part, but also incorporate the information of unoccupied orbitals (i.e., the second-order perturbative correlation) in the correlation part. Different types of DHDFs have been proposed according to different philosophies. They could be empirical as multi-coefficient methods to allow the mixing of wavefunction-based methods with the hybrid density functional methods in order to achieve a good compromise of accuracy, cost, and ease of use for practical calculations, or they have their roots in multideterminant extension of the Kohn-Sham scheme or Görling–Levy's coupling-constant perturbative theory. In this chapter, we first introduce a classification of the current DHDFs (Sect. 2.1). We then, in Sect. 2.2, discuss the Levy constrained search approach and adiabatic connection formalism, which provide a formal route that the exchange-correlation functional can be pursued. Finally, the underlying physics for the B2PLYP-type DHDFs and the XYG3-type DHDFs is explored in Sects. 2.3 and 2.4, respectively.

Keywords Levy constrained search approach · Adiabatic connection formalism · Görling–Levy's coupling-constant perturbative theory · MP2 · Multi-coefficient method · Doubly hybrid density functional

2.1 Classification of Current DHDFs

While the fourth rung functionals introduce nonlocality into the exchange functional by using a HF(Hartree-Fock)-like exchange E_x^{HF} (see Eq. 1.67), a simple combination of E_x^{HF} with a local or semilocal correlation functional E_c^{DFA} was unsuccessful (Eq. 1.68). There is a need to introduce nonlocality into the correlation functional so as to make a good match within E_{xc}. This calls for the fifth

I. Y. Zhang and X. Xu, *A New-Generation Density Functional*,
SpringerBriefs in Molecular Science, DOI: 10.1007/978-3-642-40421-4_2,
© The Author(s) 2014

rung functionals that include also the information of unoccupied orbitals [1, 2]. There are several ways that the fifth rung functionals can be constructed (e.g. [3–15]). Here, we focus on the so-called doubly hybrid density functionals (DHDFs).

2.1.1 The MC3BB Type

Truhlar and co-workers coined, for the first time, the word 'doubly hybrid' and proposed the MC3BB method [12], where the conventional HF total energy and a scaled MP2 (second-order Møller-Plesset) correlation energy are mixed with the DFT (density functional theory) total energy as in Eq. 2.1.

$$E_{\text{tot}}^{\text{MC3BB}} = e_2 \left(E_{\text{tot}}^{\text{HF}} + e_1 E_c^{\text{MP2}} \right) + (1 - e_2) E_{\text{tot}}^{\text{BBX}} \tag{2.1}$$

Here the DFT part, *BBX*, uses Becke88 [16] as its exchange and Becke95 [17] as its correlation, while X stands for the percentage of the HF-like exchange [18, 19]. The *BBX* functional adopts a form of one-mixing-parameter hybrid as in Eq. 1.70. The MP2 correlation energy is calculated using the HF orbitals $\{\phi_{i,a}\}$ with eigenvalues $\{\varepsilon_{i,a}\}$,

$$E_c^{\text{MP2}} = \frac{1}{4} \sum_{ij}^{occ} \sum_{ab}^{vir} \frac{\left| \langle \phi_i \phi_j || \phi_a \phi_b \rangle \right|^2}{\varepsilon_i + \varepsilon_j - \varepsilon_a - \varepsilon_b} \tag{2.2}$$

where the subscripts (i, j) and (a, b) denote the occupied and virtual (unoccupied) HF orbitals, respectively. $\langle \phi_i \phi_j || \phi_a \phi_b \rangle = \langle \phi_i \phi_j | \phi_a \phi_b \rangle - \langle \phi_i \phi_j | \phi_b \phi_a \rangle$ is an antisymmetrized combination of the regular two-electron repulsion integrals

$$\langle \phi_i \phi_j | \phi_a \phi_b \rangle = \iint \phi_i^*(\vec{r}_1) \phi_j^*(\vec{r}_2) r_{12}^{-1} \phi_a(\vec{r}_1) \phi_b(\vec{r}_2) \mathrm{d}\vec{r}_1 \mathrm{d}\vec{r}_2 \tag{2.3}$$

The parameters $\{e_1, e_2, X\} = \{1.332, 0.205, 0.39\}$ were obtained by fitting against a set of 109 atomization energies and 42 barrier heights. These parameters are basis set specific. In particular, the MP2 correlation energy and the HF total energy in Eq. 2.1 should be evaluated at the basis set level of $6-31 + G(d,p)$ within the frozen core approximation as originally designed [12]. The recommended basis set for the BBX calculations is an augmented polarized triple-zeta basis set MG3S, which is the same as $6\text{-}311 + G(3d2f,2df,2p)$ for H-Si but improved for P-Ar and no diffuse functions on hydrogens [20, 21].

It has to be noted that while the DFT part in Eq. 2.1 uses the KS (Kohn-Sham) orbitals based on the xc functional of $E_{\text{xc}}^{\text{BBX}}$, the MP2 part uses the HF orbitals. Therefore, there are two sets of densities associated with Eq. 2.1, which goes beyond the frame work of KS DFT [22]. By assuming that the HF density and orbitals are the same as those from DFT, one may view the multicoefficient method

MC3BB in terms of exchange-correlation functional, i.e., change the expression for
total energy as in Eq. 2.1 to that for exchange-correlation functional [23]

$$E_{xc}^{MC3BB}[\rho] = f_1 E_x^{HF} + (1 - f_1)E_x^S + f_2 \Delta E_x^B + (1 - e_2)E_c^{B95} + e_1 e_2 E_c^{MP2} \quad (2.4)$$

where $f_1 = e_2 + (1 - e_2)X = 0.515$, and $f_2 = (1 - f_1) = 0.485$. From Eq. 2.4, it
can be seen that, in addition to the hybridization in the exchange part
$(E_x^{HF}, E_x^S, \Delta E_x^B)$, the correlation part is also a hybrid between E_c^{B95} and E_c^{MP2}, i.e.,
doubly hybrid.

The name MC3 suggests that this is a multicoefficient method that contains
three parameters. It is empirical with the purpose to generalize the originally
wavefunction based multicoefficient methods to allow the mixing with the hybrid
density functional methods in order to achieve a good compromise of accuracy,
cost, and ease of use for practical calculations [12].

2.1.2 The B2PLYP Type

Grimme proposed a widely recognized DHDF B2PLYP [13]. It employs two
parameters: one is to hybridize the HF exchange with the Becke88 exchange
functional, while the other is to hybridize the MP2-like correlation with the LYP
correlation functional [24].

$$E_{xc}^{B2PLYP} = a_x E_x^{HF} + (1 - a_x)(E_x^S + \Delta E_x^B) + a_c E_c^{LYP} + (1 - a_c)E_c^{MP2} \quad (2.5)$$

The parameters $\{a_x, a_c\} = \{0.53, 0.73\}$ [13] were determined by a parameterization against heats of formation (HOFs) of the G2/97 set [25, 26].

Note that the orbitals and orbital eigenvalues used to evaluate each term,
including E_c^{MP2}, in Eq. 2.5 are from the self-consistent-filed (SCF) calculation
based on the ansatz of the DFT part alone in B2PLYP. Therefore, Eq. 2.5 can be
reformulated as

$$E_{xc}^{B2PLYP} = E_{xc,SCF}^{DFA} + (1 - a_c)E_c^{MP2} \quad (2.6)$$

The mere purpose of the generalized KS type of calculation with $E_{xc,SCF}^{DFA}$, similar
to the HF calculation in the standard MP2 theory, is to provide a reference state for
the followed perturbation calculation. Unlike the HF ansatz, the DFT part, $E_{xc,SCF}^{DFA}$,
used to generate the orbitals has already contained an a_c portion of the LYP
correlation. Hence the MP2-like correlation in B2PLYP is scaled by a factor of
$(1 - a_c)$. There is no intension with these orbitals to give a mimic of the ground
state density of the real system. This again goes beyond the KS frame work. The
theoretical foundation of the B2PLYP-type DHDFs was later provided by Savin
and co-workers based on the multideterminant extension of the Kohn-Sham
scheme (see Sect. 2.3.1) [15].

2.1.3 The XYG3 Type

Based on the adiabatic connection formalism [3, 18, 27–29], (see Sect. 2.2.2) and Görling–Levy coupling-constant perturbation expansion to the second order E_c^{GL2} [30] (see Sect. 2.4.2), a new type of DHDF, namely XYG3, was proposed, which takes the form as [14]:

$$E_{xc}^{XYG3}[\rho] = d_1 E_x^{HF} + (1 - d_1)E_x^S + d_2 \Delta E_x^B + (1 - d_3)E_c^{LYP} + d_3 E_c^{MP2} \qquad (2.7)$$

The parameters $\{d_1, d_2, d_3\} = \{0.8033, 0.2107, 0.3211\}$ [14] were determined by a parameterization against HOFs of the G3/99 set [25, 26, 31].

XYG3 distinguishes itself from the other DHDFs by using B3LYP orbitals and orbital eigenvalues to evaluate each term in Eq. 2.7. In such a way, XYG3 also shares with B3LYP the kinetic energy $T_s[\rho]$, the Coulomb energy $J[\rho]$, and the external potential energy $V_{ext}[\rho]$ in the construction of its total energy. As B3LYP is one of the most successful DFAs, it would be reasonable to expect that B3LYP gives good density that approximates well the true ground state density. This, we believe, holds the key to the success of the XYG3-type DH functionals [14, 32–42].

We may therefore classify the DHDFs currently available into three groups according to which orbitals are used to evaluate the perturbative correlation energy. In Truhlar's MC3BB [12], HF orbitals are used to compute the MP2 correlation energy, which is then mixed with total energy from a conventional hybrid meta-GGA as defined in Eq. 2.1. In the B2PLYP family of functionals [13], a truncated DFA method is used to generate orbitals and density, whose total energy is then augmented with a scaled MP2 term, evaluated by the as produced orbitals, to normalize with the DFT correlation. XYG3 uses B3LYP to generate density and orbitals, which provide a good approximation to the real density [14]. While the MC3BB type of functionals is purely empirical as a generalization of the multicoefficient method [12], the B2PLYP and the XYG3 types of functionals have their own theoretical bases [14, 15, 30, 37], respectively, which we will discuss in the following sections.

2.2 Fundamental Ideas Behind DHDFs

2.2.1 Levy Constrained Search Approach

We start by introducing the "constrained search" approach of Levy [43], which provides a constructive view of Hohenberg-Kohn Theorems and Kohn-Sham scheme [22, 44].

Solving the Schrödinger equation (Eq. 1.1) for ground state energy is equivalent to minimizing $\langle \Psi | \hat{H} | \Psi \rangle$ over all normalized, antisymmetric N-particle wavefunctions [43]:

$$E = \min_{\Psi \to N} \langle \Psi | \hat{H} | \Psi \rangle \tag{2.8}$$

This can be achieved in a two-step fashion:

$$E = \min_{\rho \to N} \left\{ \min_{\Psi \to \rho} \langle \Psi | \hat{H} | \Psi \rangle \right\} \tag{2.9}$$

In the first step, we consider all wavefunctions which yield a given density ρ, and in the second step, we consider all allowed densities. The minimizing density is then the ground state density ρ_0, which is just what has already been stated in Eq. 1.11 as the Hohenberg-Kohn theorems. Comparing Eqs. 2.9 and 1.11, one arrives at a definition of functional $E[\rho]$:

$$
\begin{aligned}
E[\rho] &= \min_{\Psi \to \rho} \langle \Psi | \hat{H} | \Psi \rangle \\
&= \min_{\Psi \to \rho} \langle \Psi | \hat{T} + \hat{V}_{ee} | \Psi \rangle + \int \rho(\vec{r}) v_{ext}(\vec{r}) d\vec{r}
\end{aligned}
\tag{2.10}
$$

The last term in Eq. 2.10 has used the fact that Ψ giving the same ρ also gives the same $\langle \Psi | \hat{V}_{ext} | \Psi \rangle$. Equation 2.10, in turn, as compared to Eq. 1.7, defines the Hohenberg-Kohn functional $F_{HK}[\rho]$:

$$F_{HK}[\rho] = \min_{\Psi \to \rho} \langle \Psi | \hat{T} + \hat{V}_{ee} | \Psi \rangle \tag{2.11}$$

For the Kohn-Sham system of noninteracting electrons, $\hat{V}_{ee}$ vanishes so that $F_{HK}[\rho]$ reduces to

$$T_s[\rho] = \min_{\Psi \to \rho} \langle \Psi | \hat{T} | \Psi \rangle = \min_{\Phi \to \rho} \langle \Phi | \hat{T} | \Phi \rangle \tag{2.12}$$

where the N-particle wave function Ψ is reduced to Φ which corresponds to the single Slater determinant constructed from one-electron orbitals $\{\phi_i\}$ (c.f. Eq. 1.16).

2.2.2 Adiabatic Connection Formalism

The Hohenberg-Kohn theorems merely prove the existence of $F_{HK}[\rho]$, which gives no direct guidance to the construction of the functional. The Kohn-Sham scheme pulls out the large part of the kinetic energy via $T_s[\rho]$, but still leaves the exact xc functional $E_{xc}[\rho]$ unknown. The adiabatic connection formalism [3, 18, 27–29] provides a possible route that $E_{xc}[\rho]$ can be pursued.

Let us define a Hamiltonian $\hat{H}_\lambda$ that represents a set of systems in which the electron–electron interaction is scaled [3, 18, 27–29]

$$\hat{H}_\lambda = \hat{T} + \lambda\hat{V}_{ee} + \hat{V}_\lambda \tag{2.13}$$

The one-electron potential $\hat{V}_\lambda = \sum_{i=1}^{N} v_\lambda([\rho]; \vec{r}_i)$ is also scaled simultaneously so that a prescribed density ρ is fixed independent of λ. This sets up the so-called adiabatic connection path along λ. Then $\hat{H}_{\lambda=0} = \hat{T} + \hat{V}_s$ is for the Kohn-Sham system, and $\hat{H}_{\lambda=1} = \hat{T} + \hat{V}_{ee} + \hat{V}_{ext}$ is for the real system. This suggests:

$$\hat{V}_{\lambda=0} = \hat{V}_s \text{ and } \hat{V}_{\lambda=1} = \hat{V}_{ext} \tag{2.14}$$

For any allowed Ψ, we have

$$E_\lambda = \min_{\Psi \to \rho}\langle\Psi|\hat{H}_\lambda|\Psi\rangle = \langle\Psi_\lambda^\rho|\hat{H}_\lambda|\Psi_\lambda^\rho\rangle \tag{2.15}$$

Here we use Ψ_λ^ρ to indicate a variational solution of Eq. 2.15 that yields density ρ. Clearly, Ψ_λ^ρ is also the wavefunction that minimizes the expectation value of $(\hat{T} + \lambda\hat{V}_{ee})$:

$$F_{HK}^\lambda[\rho] = \min_{\Psi \to \rho}\langle\Psi|\hat{T} + \lambda\hat{V}_{ee}|\Psi\rangle = \langle\Psi_\lambda^\rho|\hat{T} + \lambda\hat{V}_{ee}|\Psi_\lambda^\rho\rangle \tag{2.16}$$

which defines a generalized HK functional $F_{HK}^\lambda[\rho]$.

Hellman-Feynman theorem tells us [3, 45, 46]

$$\frac{dE_\lambda}{d\lambda} = \left\langle\Psi_\lambda^\rho\left|\frac{\partial\hat{H}_\lambda}{\partial\lambda}\right|\Psi_\lambda^\rho\right\rangle \tag{2.17}$$

Inserting Eq. (2.13) into Eq. (2.17) and integrating it from $\lambda = 0$ to $\lambda = 1$, we have

$$\int_0^1\left(\frac{dE_\lambda}{d\lambda}\right)d\lambda = \int_0^1\left\langle\Psi_\lambda^\rho\left|\frac{\partial(\hat{T} + \lambda\hat{V}_{ee})}{\partial\lambda}\right|\Psi_\lambda^\rho\right\rangle d\lambda + \int_0^1\left\langle\Psi_\lambda^\rho\left|\frac{\partial\hat{V}_\lambda}{\partial\lambda}\right|\Psi_\lambda^\rho\right\rangle d\lambda \tag{2.18}$$

Hence,

$$(E_{\lambda=1} - E_{\lambda=0}) = \int_0^1\langle\Psi_\lambda^\rho|\hat{V}_{ee}|\Psi_\lambda^\rho\rangle d\lambda + (V_{\lambda=1} - V_{\lambda=0}) \tag{2.19}$$

In getting the second term on the right-hand side of Eq. 2.19, we have taken advantage of the assumption that the electron density is held fixed along the adiabatic path from $\lambda = 0$ to $\lambda = 1$. Inserting Eqs. 1.7, 1.18, 1.20, and 2.14 into Eq. 2.19, we have

$$E_{xc}[\rho] = \int_0^1\langle\Psi_\lambda^\rho|\hat{V}_{ee}|\Psi_\lambda^\rho\rangle d\lambda - J[\rho] \tag{2.20}$$

This actually defines the exchange-correlation energy in terms of the coupling-constant integration [3, 27–29, 47–49]

$$E_{xc}[\rho] = \int_0^1 W_\lambda[\rho] \mathrm{d}\lambda \tag{2.21}$$

where

$$W_\lambda[\rho] = \left\langle \Psi_\lambda^\rho \middle| \hat{V}_{ee} \middle| \Psi_\lambda^\rho \right\rangle - J[\rho] \tag{2.22}$$

From Eq. 2.22, it is immediate to see that $W_0[\rho]$ is nothing but the exchange energy of the KS determinant

$$W_0[\rho] = \left\langle \Phi^\rho \middle| \hat{V}_{ee} \middle| \Phi^\rho \right\rangle - J[\rho] = E_x^{HF} \tag{2.23}$$

2.3 Rationale of DHDFs of the B2PLYP Type

2.3.1 Multideterminant Extension of the Kohn-Sham Scheme

We start by reformulating the total energy E_{tot} for the physical system, given in Eq. 1.7, in terms of the generalized HK functional

$$E[\rho] = F_{HK}^\lambda[\rho] + V_{ext}[\rho] + \bar{E}_{Hxc}^\lambda[\rho] \tag{2.24}$$

$\bar{E}_{Hxc}^\lambda[\rho] = F_{HK}[\rho] - F_{HK}^\lambda[\rho]$ is the complement HK functional [43, 50, 51], which can be further developed in related to the KS system:

$$\bar{E}_{Hxc}^\lambda[\rho] = \left(F_{HK}[\rho] - T_s[\rho]\right) - \left(F_{HK}^\lambda[\rho] - T_s[\rho]\right) \tag{2.25}$$

where the first term in the right-hand side is $J[\rho] + E_{xc}[\rho]$ (see Eq. 1.20), while the second term is given by $\left(F_{HK}^\lambda[\rho] - T_s[\rho]\right) = \left\langle \Psi_\lambda^\rho \middle| \hat{T} + \lambda \hat{V}_{ee} \middle| \Psi_\lambda^\rho \right\rangle - \left\langle \Phi^\rho \middle| \hat{T} \middle| \Phi^\rho \right\rangle$. Hence [43, 50, 51]

$$\bar{E}_{Hxc}^\lambda[\rho] = (1 - \lambda)J[\rho] + (1 - \lambda)E_x[\rho] + \left(E_c[\rho] - E_c^\lambda[\rho]\right) \tag{2.26}$$

Here $E_c^\lambda[\rho]$ is defined as [43, 46]

$$E_c^\lambda[\rho] = \left\langle \Psi_\lambda^\rho \middle| \hat{T} + \lambda \hat{V}_{ee} \middle| \Psi_\lambda^\rho \right\rangle - \left\langle \Phi^\rho \middle| \hat{T} + \lambda \hat{V}_{ee} \middle| \Phi^\rho \right\rangle \tag{2.27}$$

which corresponds to the correlation energy of a partial interacting system. Under uniform scaling of the density $\left[\rho_\gamma(\vec{r}) = \gamma^3 \rho(\gamma \vec{r})\right], \gamma = 1/\lambda$ [46, 52–55],

$$E_c^\lambda[\rho] = \lambda^2 E_c\left[\rho_{1/\lambda}\right] \tag{2.28}$$

$$E_c^\lambda[\rho] \approx \lambda^2 E_c[\rho] \tag{2.29}$$

Equation 2.29 is an approximation where density scaling is neglected $E_c\left[\rho_{1/\lambda}\right] \approx E_c[\rho]$ [15]. $\bar{E}_{\text{Hxc}}^\lambda[\rho]$ is also called the complement λ-dependent Hartree-exchange-correlation density functional [15].

Equation 2.24 requires to carry out minimization over all allowed Ψ, which is, however, impractical, due to the existence of $\hat{V}_{\text{ee}}$ that leads to Ψ of a general multideterminant character [15, 43]:

$$E_{\text{tot}} = \min_{\Psi \to \rho}\langle\Psi|\hat{T} + \lambda\hat{V}_{\text{ee}}|\Psi\rangle + V_{\text{ext}}[\rho] + \bar{E}_{\text{Hxc}}^\lambda[\rho] \tag{2.30}$$

A density-scaled one-parameter hybrid (DS1H) approximation [15] is then defined by restricting the minimization in Eq. 2.30 to single-determinant wavefunction Φ:

$$E^{\text{DS1H},\lambda} = \min_{\Phi}\langle\Phi|\hat{T} + \lambda\hat{V}_{\text{ee}}|\Phi\rangle + V_{\text{ext}}[\rho_\Phi] + \bar{E}_{\text{Hxc}}^\lambda[\rho_\Phi] \tag{2.31}$$

This is equivalent to solving the HF problem. The single-particle equations are:

$$\begin{aligned}
\left[-\tfrac{1}{2}\nabla^2 + v_{\text{ext}}(\vec{r}) + v_J(\vec{r}) + \lambda v_x^{\text{HF}}(\vec{r}) + (1-\lambda)v_x(\vec{r})\right.\\
\left.+v_c(\vec{r}) - \tfrac{\delta E_c^\lambda[\rho]}{\delta\rho(\vec{r})}\right]\phi_i^\lambda(\vec{r}) = \varepsilon_i^\lambda\phi_i^\lambda(\vec{r})
\end{aligned} \tag{2.32}$$

If Eq. 2.29 is adopted, Eq. 2.32 is simplified as

$$\begin{aligned}
\left[-\tfrac{1}{2}\nabla^2 + v_{\text{ext}}(\vec{r}) + v_J(\vec{r}) + \lambda v_x^{\text{HF}}(\vec{r}) + (1-\lambda)v_x(\vec{r})\right.\\
\left.+(1-\lambda^2)v_c(\vec{r})\right]\phi_i^\lambda(\vec{r}) = \varepsilon_i^\lambda\phi_i^\lambda(\vec{r})
\end{aligned} \tag{2.33}$$

which corresponds to the so-called 1H approximation without considering density scaling [15]. The final DS1H and 1H energies are then given, respectively, by

$$\begin{aligned}
E_{\text{SCF}}^{\text{DS1H},\lambda} = -\frac{1}{2}\sum_{i=1}^{N}\int\left(\phi_i^\lambda\right)^*\nabla^2\phi_i^\lambda d\vec{r} + V_{\text{ext}}\left[\rho_{\Phi^\lambda}\right] + J\left[\rho_{\Phi^\lambda}\right] + \lambda E_x^{HF}\left[\rho_{\Phi^\lambda}\right]\\
+ (1-\lambda)E_x\left[\rho_{\Phi^\lambda}\right] + E_c\left[\rho_{\Phi^\lambda}\right] - E_c^\lambda\left[\rho_{\Phi^\lambda}\right]
\end{aligned} \tag{2.34}$$

$$\begin{aligned}
E_{\text{SCF}}^{\text{1H},\lambda} = -\frac{1}{2}\sum_{i=1}^{N}\int\left(\phi_i^\lambda\right)^*\nabla^2\phi_i^\lambda d\vec{r} + V_{\text{ext}}\left[\rho_{\Phi^\lambda}\right] + J\left[\rho_{\Phi^\lambda}\right] + \lambda E_x^{HF}\left[\rho_{\Phi^\lambda}\right]\\
+ (1-\lambda)E_x\left[\rho_{\Phi^\lambda}\right] + (1-\lambda^2)E_c\left[\rho_{\Phi^\lambda}\right]
\end{aligned} \tag{2.35}$$

Equation 2.35 has a similar form for its E_{xc} as that in the standard one-parameter hybrid functionals (c.f. Eq. 1.70) such as B1LYP or PBE0 [28, 56]. One salient difference, however, is that the correlation energy is weighted by $(1-\lambda^2)$

while in the standard one-parameter hybrid functionals it is weighted by a factor of 1. Due to the scaled interaction $\lambda \hat{V}_{ee}$, only partial correlation energy is contained in Eqs. 2.34 and 2.35. The missing part can be repaired by a nonlinear Rayleigh-Schrödinger perturbation theory [57] staring from the DS1H or 1H reference. Just like in standard Moller-Plesset perturbation theory, the final energy up to second order is given by [15]:

$$E^{\text{DS1DH},\lambda} = E_{\text{SCF}}^{\text{DS1H},\lambda} + \lambda^2 E_c^{\text{MP2}} \qquad (2.36)$$

$$E^{\text{1DH},\lambda} = E_{\text{SCF}}^{\text{1H},\lambda} + \lambda^2 E_c^{\text{MP2}} \qquad (2.37)$$

This provides a rationale for DHDFs of the B2PLYP type. One has to note that, just like the HF theory whose density is by definition not the ground state density of the real system, ρ_{Φ^λ} by construction is not meant to be the ground state density of the real system. From Eq. 2.30 to 2.31, one starts from the adiabatic connection formulism, but eventually departs from the adiabatic connection path by replacing the general multideterminant Ψ with a single-determinant wavefunction Φ with no constraint on the density ρ.

2.3.2 Development of DHDFs of the B2PLYP Type

B2PLYP was proposed before DS1DH and 1DH. The later explored the theoretical foundation of the B2PLYP-type DHDFs in terms of multideterminant extension of the Kohn-Sham scheme. On the other hand, as Eq. 2.29 is just an approximation, the scaling factor before $E_c[\rho]$ needs not to be λ^2. Hence $\{a_x, a_c\}$ in B2PLYP can be optimized independently. In fact, B2PLYP was originally understood as an interpolation approach between pure GGA-DFT and MP2, respectively [13]. As seen from Eq. 2.5, if $\{a_x, a_c\} = \{1.0, 0.0\}$, MP2 is recovered, while if $\{a_x, a_c\} = \{0.0, 1.0\}$, B2PLYP is reduced to BLYP. Furthermore, if $\{a_x, a_c\} = \{0.0, 0.0\}$ (i.e., B-MP2), an exchange-only SCF calculation may be first performed, whose orbitals and orbital eigenvalues are then used to get the MP2 energy. If $\{a_x, a_c\} = \{1.0, 1.0\}$ (i.e., HF-LYP), the full portion of nonlocal HF exchange is in company with the semilocal LYP correlation energy.

As typical DFT correlation functionals are superior to MP2 in the description of short-range correlation, and MP2 is very well suited for the description of long-range correlation, it was therefore expected that the doubly hybrid functionals that combine the two should handle both types of correlation better than either conventional DFTs and MP2 [13]. On the other hand, there are areas characterized by some physically inappropriate combination of $\{a_x, a_c\}$ which leads to, e.g., unbalanced treatment of dynamical and static correlation effects [13, 58]. Hence, $\{a_x, a_c\}$ should be optimized.

Along this line, several new functionals of the B2PLYP type (e.g., B2T-PLYP [59], B2K-PLYP [58, 59], B2GP-PLYP [58], B2π-PLYP [23], ROB2-PLYP [60],

UB2-PLYP [60], ωB97X-2 [61] etc.) have been developed. While $\{a_x, a_c\}$ in B2T-PLYP are optimized against some thermodynamic data, those in B2K-PLYP are optimized against some kinetic data [59]. For a chemical reaction, the forward and backward reaction barrier heights are connected through the reaction thermodynamics. Hence, there is a need for some functionals to be of general purpose (e.g., B2GP-PLYP) [58]. The B2π-PLYP functional was optimized specially for π-conjugated systems [23]. ROB2-PLYP and UB2PLYP recognize the importance to distinguish a restricted and an unrestricted calculation for open shell systems [60]. The ωB97X-2 [61] functional can also be put into the B2PLYP class where the truncated DFT is a long-range corrected hybrid of the B97 [62] type.

Figure 2.1 illustrates the area of $\{a_x, a_c\}$ where the B2PLYP family of functionals based on BLYP work best for certain properties. More recently, other GGA and meta-GGA functionals have come into play [13, 59, 64]. Not only the mixing parameters, but also the parameters within GGAs and meta-GGAs are refitted to optimize the final performance of DHDFs [64]. These GGAs or meta-GGAs cannot be used alone, as their mere purpose is to provide a reference state where the scaled MP2 correction can be evaluated. Furthermore, as the MP2 term contains the same-spin and opposite spin components, spin-component scaled DHDFs have been introduced [65, 66].

Conventional DFAs miss the R^{-6} decay behavior in the long-range correlation [67], and hence fail badly for dispersion-dominant nonbonded interactions. The delocalized E_c^{MP2} captures the correct long-range behavior by construction. However, a global fraction ($\sim 30\%$) of the MP2 correlation is incomplete to describe the full correlation effect in the long range [68]. It was therefore suggested to add the functionals a posterior force field (FF) like dispersion correction (-D or -D3) [69–71] to impose the correct long-range R^{-6} interatomic dependence. More recently, the DFT-D or DFT-D3 method was made an integrated method where the parameters in the corresponding DFA (i.e., electronic) part were

Fig. 2.1 The B2PLYP family of functionals based on BLYP [76]

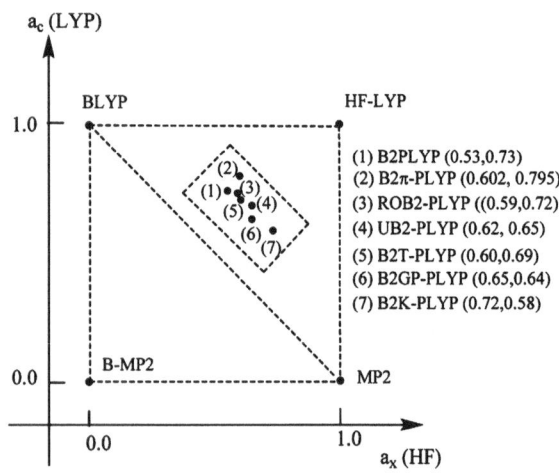

optimized in conjunction with the optimization of the parameters in the FF-like dispersion correction terms, in a hope that the medium- to long-range correlations were completely took over by the classic FF terms, and not mimicked by the DFA part to avoid double-counting [64, 66]. While such DFT-D or -D3 scheme can be implemented efficiently without additional computational cost, it includes many empirical parameters, and suffers from some inherent limitations [69, 72]. Especially, the many-body correlation effects and anisotropy effects in the long-range dispersive interactions, as well as the orbital-dependence in the medium range [72, 73], are more subtle, which are difficult to be approximated in the pair-wise additive FF models.

Detailed benchmark calculations of the B2PLYP-type functionals can be found in the literature [64]. Some of the key findings are represented in Chap. 3.

2.4 Rationale of DHDFs of the XYG3 Type

2.4.1 Becke's Hybrid-DFT Methods

In Sect. 1.2.5, we have tried to motivate Becke's hybrid-DFT methods from a view for a balanced treatment of static and dynamic correlations [74]. In fact, adiabatic connection formalism [3, 18, 27–29] provides a better way to comprehend these methods [18, 19, 75].

It is true that the main problem or challenge of the adiabatic connection approach for functional construction by using Eq. 2.21 is that the exact integrand $W_\lambda[\rho]$ is unknown. The optimistic picture is that while $\Psi(\vec{r}_1\sigma_1, \ldots, \vec{r}_N\sigma_N)$ to be solved in the Schrödinger equation is of $4N$ dimension, the divine xc functional $E_{xc}[\rho]$ characterized by ρ is only 3-dimensional. Hence, the problem is simplified. Within the adiabatic connection approach, one can then focus on the λ space of 1-dimension to approximate $W_\lambda[\rho]$ by a model function of λ. This may seem to be an even 'easier' task. In fact, this scenario has been explored for functional construction by several researchers [18, 19, 28, 29].

Becke assumed a linear model for $W_\lambda[\rho]$ [18]:

$$W_\lambda[\rho] = a[\rho] + b[\rho]\lambda \tag{2.38}$$

Integrating Eq. 2.21 by using Eq. 2.38 leads to an energy expression

$$E_{xc}[\rho] = a + \frac{1}{2}b \tag{2.39}$$

From Eq. 2.38, it is obvious that

$$a = W_0[\rho]; \; b = W_1[\rho] - W_0[\rho] \tag{2.40}$$

While $W_0[\rho]$ is just the HF exchange (Eq. 2.23), $W_1[\rho]$ corresponds to the potential energy contribution to the xc energy of the fully interacting system, which is yet unknown. Originally, Becke used a LDA to approximate $W_1[\rho]$ [18]:

$$W_1[\rho] \approx W_{xc}^{LDA} = E_x^{LDA} + W_c^{LDA} \qquad (2.41)$$

Nevertheless, it is quite common to approximate W_c^{LDA} with E_c^{LDA} or $\frac{1}{2}W_c^{LDA}$ with E_c^{GGA} such as E_c^{LYP}, leading to

$$E_{xc}^{BHandH}[\rho] = \frac{1}{2}\left(E_x^{HF} + E_x^S\right) + \frac{1}{2}E_c^{LYP} \qquad (2.42)$$

$$E_{xc}^{BHandHLYP}[\rho] = \frac{1}{2}\left(E_x^{HF} + E_x^B\right) + E_c^{LYP} \qquad (2.43)$$

Equation 2.42 corresponds to the so-called BHandH functional, while Eq. 2.43 stands for the so-called BHandHLYP functional as implemented in Gaussian suite of program [76]. Mixing in some portion of the HF exchange results in a significant improvement in functional performance for covalent bonded systems over (semi)local functionals. The take-home message from Eq. 2.43 is that the source of errors shall be traced back to the improper behavior of (semi) local exchange functionals at the $\lambda \rightarrow 0$ limit [18]. Within this linear model, half of the (semi)-local exchange should be replaced by half of the HF exchange.

For greater accuracy, Becke introduced the well-established three-parameter scheme (B3, Eq. 1.71) [19], which is a more empirical model by relaxing the linear approximation. Heavily parameterized functionals (e.g., the M06 family of functionals [77, 78]) have appeared recently, which have push limits of the hybrid functionals, providing very high accuracy for a broad range of complex systems.

2.4.2 Coupling-Constant Expansion

Let us start by reformulating the Hamiltonian $\hat{H}_\lambda$ (Eq. 2.13) of the partially interacting system with respect to the Hamiltonian $\hat{H}_s$ (1.14) of the noninteracting KS system

$$\begin{aligned}\hat{H}_\lambda &= \hat{T} + \lambda\hat{V}_{ee} + \sum_{i=1}^N v_\lambda([\rho]; \vec{r}_i) \\ &= \left(\hat{T} + \sum_{i=1}^N v_s([\rho]; \vec{r}_i)\right) + \lambda H'\end{aligned} \qquad (2.44)$$

In writing Eq. 2.44, we have emphasized the fact that density is fixed to be ρ for any $\hat{H}_\lambda$ of varying λ. Equation 2.44 defines the perturbation

$$\hat{H}' = \hat{V}_{ee} + \frac{1}{\lambda} \sum_{i=1}^{N} [v_\lambda(\vec{r}_i) - v_s(\vec{r}_i)] \tag{2.45}$$

According to the standard perturbation theory [30, 54], the ground state energy associated with $\hat{H}_\lambda$ can be written down as:

$$E_\lambda = E_s^0 + \lambda E^{(1)} + \lambda^2 E^{(2)} + \lambda^3 E^{(3)} + O(\lambda^4) \tag{2.46}$$

where E_s^0 corresponds to the ground state energy associated with $\hat{H}_s$, and $E^{(k)}$ refers to the kth-order energy correction to E_s^0. Here, we have assumed that $\hat{H}_s$ has a nondegenerate ground state $\Phi_s^0 = \Phi^\rho$ with other eigenstates Φ_s^m of eigenvalues $E_s^m (m \neq 0)$. Hence,

$$E_s^0 = \langle \Phi_s^0 | \hat{H}_s | \Phi_s^0 \rangle \tag{2.47}$$

$$E^{(1)} = \langle \Phi_s^0 | \hat{H}' | \Phi_s^0 \rangle \tag{2.48}$$

$$E^{(2)} = \sum_{m \neq 0}^{\infty} \frac{|\langle \Psi_s^0 | \hat{H}' | \Psi_s^m \rangle|^2}{E_s^0 - E_s^m} \tag{2.49}$$

This provides a tool that correlation energy with scaled density may be estimated [30, 43, 46]:

$$
\begin{aligned}
E_c\left[\rho_{1/\lambda}\right] &= \frac{1}{\lambda^2} E_c^\lambda[\rho] = \frac{1}{\lambda^2} \left[\left\langle \Psi_\rho^\lambda \middle| \hat{T} + \lambda \hat{V}_{ee} \middle| \Psi_\rho^\lambda \right\rangle - \left\langle \Phi_s^0 \middle| \hat{T} + \lambda \hat{V}_{ee} \middle| \Phi_s^0 \right\rangle \right] \\
&= \frac{1}{\lambda^2} \left[\left\langle \Psi_\rho^\lambda \middle| \hat{H}_\lambda \middle| \Psi_\rho^\lambda \right\rangle - \left\langle \Phi_s^0 \middle| \hat{H}_s \middle| \Phi_s^0 \right\rangle - \lambda \left\langle \Phi_s^0 \middle| \hat{H}' \middle| \Phi_s^0 \right\rangle \right] \\
&= E^{(2)} + \lambda O(\lambda^3)
\end{aligned}
\tag{2.50}
$$

In order to keep the density ρ constant for all values of the coupling constant $\lambda = 0 \to 1$, it is required that the chemical potential μ should be constant for this family of partially interacting N-electron systems. As the generalized HK functional is given by

$$
\begin{aligned}
F_{HK}^\lambda[\rho] &= F_{HK}[\rho] - \bar{E}_{Hxc}^\lambda[\rho] \\
&= T_s[\rho] + \lambda J[\rho] + \lambda E_x[\rho] + \lambda^2 E_c\left[\rho_{1/\lambda}\right]
\end{aligned}
\tag{2.51}
$$

the corresponding Euler-Lagrange equation (c.f. Eqs. 1.13 and 1.21) becomes

$$\mu = \frac{\delta F_{HK}^\lambda[\rho]}{\delta \rho(\vec{r})} + v_\lambda(\vec{r}) = \frac{\delta T_s[\rho]}{\delta \rho(\vec{r})} + v_s(\vec{r}) \tag{2.52}$$

giving

$$\left(v_\lambda(\vec{r}_i) - v_s(\vec{r}_i)\right) = -\lambda\left[v_J(\vec{r}_i) + v_x(\vec{r}_i) + \lambda\frac{\delta E_c\left[\rho_{1/\lambda}\right]}{\delta\rho(\vec{r}_i)}\right] \qquad (2.53)$$

Hence $\hat{H}'$ in Eq. 2.45 can be simplified for $\lambda \to 0$ as [30]:

$$\Delta = \lim_{\lambda\to 0}\hat{H}' = \hat{V}_{ee} - \sum_{i=1}^{N}\left[v_J(\vec{r}_i) + v_x(\vec{r}_i)\right] \qquad (2.54)$$

Inserting Eq. 2.54 into Eq. 2.50, we arrive at a perturbation description of $E_c\left[\rho_{1/\lambda}\right]$:

$$\lim_{\lambda\to 0}E_c\left[\rho_{1/\lambda}\right] = E^{(2)} \qquad (2.55)$$

where as a weak perturbation, Eq. 2.49 is simplified as

$$E^{(2)} = \sum_{m\neq 0}^{\infty}\frac{\left|\langle\Phi_s^0|\Delta|\Phi_s^m\rangle\right|^2}{E_s^0 - E_s^m} \qquad (2.56)$$

$E^{(2)}$ is widely recognized as the Görling–Levy theory of coupling-constant perturbation expansion to the second order, which may be explicitly written in terms of KS orbitals as [30]

$$E^{(2)} = E_c^{GL2} = \sum_i^{occ}\sum_a^{vir}\frac{\left|\langle\phi_i|v_x - v_x^{HF}|\phi_a\rangle\right|^2}{\varepsilon_i - \varepsilon_a} + \frac{1}{4}\sum_{ij}^{occ}\sum_{ab}^{vir}\frac{\left|\langle\phi_i\phi_j||\phi_a\phi_b\rangle\right|^2}{\varepsilon_i + \varepsilon_j - \varepsilon_a - \varepsilon_b} \qquad (2.57)$$

Equation 2.57 should be compared with Eq. 2.2 for MP2. In addition to the difference in the meaning of orbitals, i.e., HF orbitals in MP2 and KS orbitals in GL2, there is an additional term from singles' contributions [30].

2.4.3 Development of DHDFs of the XYG3 Type

Based on the adiabatic connection formalism [3, 18, 27–29] and Görling–Levy coupling-constant perturbation expansion [30] to the second order, we proposed XYG3 [14].

For a partially interacting system, the xc functional may be defined as

$$E_{xc}^\lambda[\rho] = \langle\Psi_\lambda^\rho|\hat{T} + \lambda\hat{V}_{ee}|\Psi_\lambda^\rho\rangle - \langle\Phi^\rho|\hat{T}|\Phi^\rho\rangle - \lambda J[\rho] \qquad (2.58)$$

which can be reformulated as

$$E_{xc}^{\lambda}[\rho] = \lambda E_x^{HF}[\rho] + \lambda^2 E_c\left[\rho_{1/\lambda}\right]$$

$$= \lambda E_x^{HF}[\rho] + \lambda^2 E^{(2)} + O(\lambda^3) \qquad (2.59)$$

$$= \int_0^{\lambda} W_{\lambda'}[\rho]d\lambda'$$

Hence, for weak perturbation, we have

$$W_{\lambda}[\rho] = E_x^{HF}[\rho] + 2\lambda E_c^{GL2} + O(\lambda^2) \qquad (2.60)$$

This demonstrates that E_c^{GL2} defines the initial slope of the xc potential energy:

$$W_0' = \left.\frac{\partial W_{\lambda}}{\partial \lambda}\right|_{\lambda=0} = 2E_c^{GL2} \qquad (2.61)$$

Equation 2.60 provides a formula with which $W_{\lambda}[\rho]$ can be approximated. As compared with Eq. 2.38 for the linear model (Fig. 2.2), it is clear, instead of choosing the initial point and the ending point of $\{W_0, W_1\}$ but approximating $W_1 \approx W_{xc}^{LDA}$ as Becke did in his half and half functional [18], parameters $\{a, b\}$ in Eq. 2.39 can be rigorously fixed using $\{W_0, W_0'\}$ [14]:

$$a = E_x^{HF}; \; b = 2E_c^{GL2} \qquad (2.62)$$

This provides an exact functional in terms of KS orbitals, if the linear dependence of $W_{\lambda}[\rho]$ on λ is faithfully fulfilled.

$$E_{xc}^{linearAC} = E_x^{HF} + E_c^{GL2} \qquad (2.63)$$

Equation 2.63 may look similar to the standard MP2 theory where the HF exchange is augmented with the MP2 correlation. It has to be emphasized that the MP2 method is just the lowest level correlated wavefunction method for many electron systems, whereas Eq. 2.63 is exact for any system with linear λ-dependence.

Fig. 2.2 Linear (**a**) or nonlinear (**b**) model for adiabatic connection path [79]

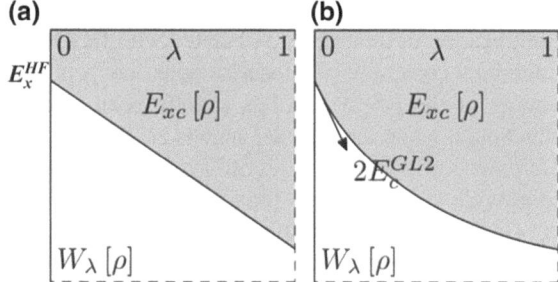

To go beyond the linear model, we proposed to choose b as in Eq. 2.64, where $\{b_1, b_2\}$ are empirical parameters introduced here to optimize the functional performance [14]:

$$b = b_1 E_c^{GL2} + b_2\left(E_{xc}^{DFT} - E_x^{HF}\right) \tag{2.64}$$

Assuming that $E_c^{DFA} \approx \left(E_{xc}^{DFA} - E_x^{HF}\right)$ contains a complete description of correlation effects, the second term of Eq. 2.64 may also be viewed as a way to extrapolate the second-order perturbation to infinite order to account for the higher order, $O(\lambda^2)$ in Eq. 2.60 dependence of W_λ on λ.

In analogy to the B3 scheme shown in Eq. 1.71, we proposed a new type of DHDFs which read as [14, 37]

$$\begin{aligned} E_{xc}^{xDH}[\rho] &= a_0 E_x^{HF} + (1 - a_0)E_x^{LDA} + a_x\Delta E_x^{GGA} \\ &+ a_1 E_c^{GL2} + (1 - a_1)E_c^{LDA} + a_c\Delta E_c^{GGA} \end{aligned} \tag{2.65}$$

Here $E_x^{GGA}[\rho] = E_x^{LDA} + \Delta E_x^{GGA}, E_c^{GGA}[\rho] = E_c^{LDA} + \Delta E_c^{GGA}$. Of course, the GGA functional can be replaced by a meta-GGA functional. Equation 2.65 suggests that not only some portion of the (semi) local exchange functional be substituted by the HF exchange, but also some portion of the (semi) local correlation functional be substituted by the GL2 correlation. While it is important to properly describe the initial value of W_λ with E_x^{HF}, it is also essential to accurately describe the initial slope of W_λ with E_c^{GL2}.

In setting up the first xDH functional XYG3 (Eq. 2.7), we have adopted LDA = SVWN [80, 81] and GGA = BLYP [16, 24]. As the LYP functional does not reduce to the correct limit for a uniform electron gas by construction, we have constrained $(1 - a_1) = a_c$, which practically eliminated one fitting parameter.

Note that single-excitation contribution is also not explicitly calculated in XYG3, such that the GL2 term [30] is actually approximated by the MP2-like PT2 term. We have argued that the singles contribution can be absorbed into E_c^{DFA} and the fitting parameters against the experimental data [14].

Unlike in LDAs and GGAs, where the xc energy is represented by an explicit functional of the density, DHDFs are formally orbital-dependent functionals which are implicit functionals of the electron density [12, 14]. The so-called optimized effective potential (OEP) method [82–86] should be invoked where the variational optimization of the energy associated with DHDFs should be carried out under the additional constraint of a local xc potential. It is interesting to note that the kinetic energy of the KS system has already been represented by an orbital-dependent functional, while the HF exchange is another evident orbital-dependent functional. Here, we again make a detour to avoid OEP by using the well-established B3LYP to provide the orbital information required for the construction of energy terms in XYG3. We believe that other fully functionalized DFAs (i.e., nontruncated) can play a similar role as B3LYP does in XYG3. In particular, pure GGAs with local multiplicative potentials are in better accordance with Görling–Levy coupling-constant perturbation expansion theory [30].

References

1. Perdew JP, Ruzsinszky A, Tao JM et al (2005) Prescription for the design and selection of density functional approximations: More constraint satisfaction with fewer fits. J Chem Phys 123:062201. doi:10.1063/1.1904565
2. Furche F, Perdew JP (2006) The performance of semilocal and hybrid density functionals in 3d transition-metal chemistry. J Chem Phys 124:044103. doi:10.1063/1.2162161
3. Langreth DC, Perdew JP (1977) Exchange-correlation energy of a metallic surface: Wave-vector analysis. Phys Rev B 15:2884–2901. doi:10.1103/PhysRevB.15.2884
4. Langreth DC, Perdew JP (1980) Theory of nonuniform electronic systems. I. Analysis of the gradient approximation and a generalization that works. Phys Rev B 21:5469–5493. doi:10.1103/PhysRevB.21.5469
5. Furche F (2001) Molecular tests of the random phase approximation to the exchange-correlation energy functional. Phys Rev B 64:195120–195128. doi:10.1103/PhysRevB.64.195120
6. Grüneis A, Marsman M, Harl J et al (2009) Making the random phase approximation to electronic correlation accurate. J Chem Phys 131:154115. doi:10.1063/1.3250347
7. Ren X, Tkatchenko A, Rinke P, Scheffler M (2011) Beyond the random-phase approximation for the electron correlation energy: the importance of single excitations. Phys Rev Lett 106:153003–153004. doi:10.1103/PhysRevLett.106.153003
8. Lie GC, Clementi E (1974) Study of the electronic structure of molecules. XXI. Correlation energy corrections as a functional of the Hartree-Fock density and its application to the hydrides of the second row atoms. J Chem Phys 60:1275–1287. doi:10.1063/1.1681192
9. Savin A, Flad H-J (1995) Density functionals for the yukawa electron-electron interaction. Int J Quantum Chem 56:327–332. doi:10.1002/qua.560560417
10. Gräfenstein J, Cremer D (2000) The combination of density functional theory with multi-configuration methods–CAS-DFT. Chem Phys Lett 316:569–577. doi:10.1016/S0009-2614(99)01326-3
11. Sharkas K, Savin A, Jensen HJA, Toulouse J (2012) A multiconfigurational hybrid density-functional theory. J Chem Phys 137:044104. doi:10.1063/1.4733672
12. Zhao Y, Lynch BJ, Truhlar DG (2004) Doubly hybrid meta DFT: New multi-coefficient correlation and density functional methods for thermochemistry and thermochemical kinetics. J Phys Chem A 108:4786–4791. doi:10.1021/jp049253v
13. Grimme S (2006) Semiempirical hybrid density functional with perturbative second-order correlation. J Chem Phys 124:034108–034116. doi:10.1063/1.2148954
14. Zhang Y, Xu X, Goddard WA (2009) Doubly hybrid density functional for accurate descriptions of nonbond interactions, thermochemistry, and thermochemical kinetics. Proc Natl Acad Sci USA 106:4963–4968. doi:10.1073/pnas.0901093106
15. Sharkas K, Toulouse J, Savin A (2011) Double-hybrid density-functional theory made rigorous. J Chem Phys 134:064113. doi:10.1063/1.3544215
16. Becke AD (1988) Density-functional exchange-energy approximation with correct asymptotic behavior. Phys Rev A 38:3098–3100. doi:10.1103/PhysRevA.38.3098
17. Becke AD (1996) Density-functional thermochemistry. 4. A new dynamical correlation functional and implications for exact-exchange mixing. J Chem Phys 104:1040–1046. doi:10.1063/1.470829
18. Becke AD (1993) A new mixing of Hartree–Fock and local density-functional theories. J Chem Phys 98:1372–1377. doi:10.1063/1.464304
19. Becke AD (1993) Density-functional thermochemistry. 3: The role of exact exchange. J Chem Phys 98:5648–5652. doi:10.1063/1.464913
20. Fast PL, Sánchez ML, Truhlar DG (1999) Multi-coefficient Gaussian-3 method for calculating potential energy surfaces. Chem Phys Lett 306:407–410. doi:10.1016/S0009-2614(99)00493-5

21. Curtiss LA, Redfern PC, Raghavachari K et al (1999) Gaussian-3 theory using reduced Møller-Plesset order. J Chem Phys 110:4703–4709. doi:10.1063/1.478385
22. Kohn W, Sham LJ (1965) Self-consistent equations including exchange and correlation effects. Phys Rev 140:A1133–A1138. doi:10.1103/PhysRev.140.A1133
23. Sancho-García JC, Pérez-Jiménez AJ (2009) Assessment of double-hybrid energy functionals for pi-conjugated systems. J Chem Phys 131:084108–084111. doi:10.1063/1.3212881
24. Lee CT, Yang WT, Parr RG (1988) Development of the Colle–Salvetti correlation-energy formula into a functional of the electron-density. Phys Rev B 37:785–789. doi:10.1103/PhysRevB.37.785
25. Curtiss LA, Raghavachari K, Trucks GW, Pople JA (1991) Gaussian-2 theory for molecular-energies of 1st-row and 2nd-row compounds. J Chem Phys 94:7221–7230. doi:10.1063/1.460205
26. Curtiss LA, Raghavachari K, Redfern PC et al (1998) Gaussian-3 (G3) theory for molecules containing first and second-row atoms. J Chem Phys 109:7764–7776. doi:10.1063/1.477422
27. Harris J, Jones RO (1974) The surface energy of a bounded electron gas. J Phys F 4:1170–1186. doi:10.1088/0305-4608/4/8/013
28. Langreth DC, Perdew JP (1975) The exchange-correlation energy of a metallic surface. Solid State Commun 17:1425–1429. doi:10.1016/0038-1098(75)90618-3
29. Gunnarsson O, Lundqvist BI (1976) Exchange and correlation in atoms, molecules, and solids by the spin-density-functional formalism. Phys Rev B 13:4274–4298. doi:10.1103/PhysRevB.13.4274
30. Görling A, Levy M (1993) Correlation-energy functional and its hight-density limit obtained from a coupling-constant perturbation expansion. Phys Rev B 47:13105–13113. doi:10.1103/PhysRevB.47.13105
31. Curtiss LA, Raghavachari K, Redfern PC, Pople JA (2000) Assessment of Gaussian-3 and density functional theories for a larger experimental test set. J Chem Phys 112:7374–7383. doi:10.1063/1.481336
32. Zhang IY, Wu J, Luo Y, Xu X (2010) Trends in R−X bond dissociation energies (R· = Me, Et, i-Pr, t-Bu, X· = H, Me, Cl, OH). J Chem Theory Comput 6:1462–1469. doi:10.1021/ct100010d
33. Zhang IY, Luo Y, Xu X (2010) XYG3s: Speedup of the XYG3 fifth-rung density functional with scaling-all-correlation method. J Chem Phys 132:194105–194111. doi:10.1063/1.3424845
34. Zhang IY, Wu JM, Xu X (2010) Extending the reliability and applicability of B3LYP. Chem Comm 46:3057–3070. doi:10.1039/c000677g
35. Zhang IY, Wu J, Luo Y, Xu X (2011) Accurate bond dissociation enthalpies by using doubly hybrid XYG3 functional. J Comput Chem 32:1824–1838. doi:10.1002/jcc.21764
36. Zhang IY, Xu X, Jung Y, Goddard WA (2011) A fast doubly hybrid density functional method close to chemical accuracy using a local opposite spin ansatz. Proc Natl Acad Sci USA 108:19896–19900. doi:10.1073/pnas.1115123108
37. Zhang IY, Xu X (2011) Doubly hybrid density functional for accurate description of thermochemistry, thermochemical kinetics and nonbonded interactions. Int Rev Phys Chem 30:115–160. doi:10.1080/0144235X.2010.542618
38. Liu G, Wu J, Zhang IY et al (2011) Theoretical studies on thermochemistry for conversion of 5-Chloromethylfurfural into valuable chemicals. J Phys Chem A 115:13628–13641. doi:10.1021/jp207641g
39. Shen C, Zhang IY, Fu G, Xu X (2011) Pyrolysis of D-Glucose to Acrolein. Chin J Chem Phys 24:249–252. doi:10.1088/1674-0068/24/03/249-252
40. Zhang IY, Su NQ, Brémond ÉAG et al (2012) Doubly hybrid density functional xDH-PBE0 from a parameter-free global hybrid model PBE0. J Chem Phys 136:174103. doi:10.1063/1.3703893
41. Zhang IY, Xu X (2012) Gas–Phase thermodynamics as a validation of computational catalysis on surfaces: A case study of Fischer–Tropsch synthesis. ChemPhysChem 13:1486–1494. doi:10.1002/cphc.201100909

42. Zhang IY, Xu X (2012) XYG3 and XYGJ-OS performances for noncovalent binding energies relevant to biomolecular structures. Phys Chem Chem Phys 14:12554. doi:10.1039/c2cp40904f

43. Levy M (1979) Universal variational functionals of electron densities, 1st-order density matrices, and natural spin-orbitals and solution of the V-representability problem. Proc Natl Acad Sci USA 76:6062–6065. doi:10.1073/pnas.76.12.6062

44. Hohenberg P, Kohn W (1964) Inhomogeneous electron gas. Phys Rev B 136:B864–B871. doi:10.1103/PhysRev.136.B864

45. Feynman RP (1939) Forces in molecules. Phys Rev 56:340–343. doi:10.1103/PhysRev.56.340

46. Levy M, Perdew JP (1985) Hellmann-Feynman, virial, and scaling requisites for the exact universal density functionals. Shape of the correlation potential and diamagnetic susceptibility for atoms. Phys Rev A 32:2010–2021. doi:10.1103/PhysRevA.32.2010

47. Levy M (1983) Density Functional Theory. Springer, New York

48. Perdew JP, Emzerhof M, Burke K (1996) Rationale for mixing exact exchange with density functional approximations. J Chem Phys 105:9982–9985. doi:10.1063/1.472933

49. Mori-Sánchez P, Cohen AJ, Yang WT (2006) Self-interaction-free exchange-correlation functional for thermochemistry and kinetics. J Chem Phys 124:091102. doi:10.1063/1.2179072

50. Toulouse J, Colonna F, Savin A (2004) Long-range–short-range separation of the electron-electron interaction in density-functional theory. Phys Rev A 70:062505. doi:10.1103/PhysRevA.70.062505

51. Ángyán J, Gerber I, Savin A, Toulouse J (2005) van der Waals forces in density functional theory: Perturbational long-range electron-interaction corrections. Phys Rev A 72:012510–012519. doi:10.1103/PhysRevA.72.012510

52. Levy M, Yang WT, Parr RG (1985) A new functional with homogeneous coordinate scaling in density functional theory: F [ρ, λ]. J Chem Phys 83:2334–2336. doi:10.1063/1.449326

53. Levy M (1991) Density-functional exchange correlation through coordinate scaling in adiabatic connection and correlation hole. Phys Rev A 43:4637–4646. doi:10.1103/PhysRevA.43.4637

54. Szabo A, Ostlund NS (1982) Modern quantum chemistry. MacMillan, New York

55. Levy M, Perdew JP (1993) Tight bound and convexity constraint on the exchange-correlation-energy functional in the low-density limit, and other formal tests of generalized-gradient approximations. Phys Rev B 48:11638–11645. doi:10.1103/PhysRevB.48.11638

56. Adamo C, Barone V (1999) Toward reliable density functional methods without adjustable parameters: The PBE0 model. J Chem Phys 110:6158–6170. doi:10.1063/1.478522

57. Fromager E, Jensen HJA (2008) Self-consistent many-body perturbation theory in range-separated density-functional theory: A one-electron reduced-density-matrix-based formulation. Phys Rev A 78:022504. doi:10.1103/PhysRevA.78.022504

58. Karton A, Tarnopolsky A, Lamere JF et al (2008) Highly accurate first-principles benchmark data sets for the parametrization and validation of density functional and other approximate methods. Derivation of a robust, generally applicable, double-hybrid functional for thermochemistry and thermochemical kinetics. J Phys Chem A 112:12868–12886. doi:10.1021/jp801805p

59. Tarnopolsky A, Karton A, Sertchook R et al (2008) Double-hybrid functionals for thermochemical kinetics. J Phys Chem A 112:3–8. doi:10.1021/jp710179r

60. Graham D, Menon A, Goerigk L et al (2009) Optimization and basis-set dependence of a restricted-open-shell form of B2-PLYP double-hybrid density functional theory. J Phys Chem A 113:9861–9873. doi:10.1021/jp9042864

61. Chai J-D, Head-Gordon M (2009) Long-range corrected double-hybrid density functionals. J Chem Phys 131:174105. doi:10.1063/1.3244209

62. Becke AD (1997) Density-functional thermochemistry. 5. Systematic optimization of exchange-correlation functionals. J Chem Phys 107:8554–8560. doi:10.1063/1.475007

63. Zhang IY (2011) A new generation density functional towards chemical accuracy. Doctorial thesis, KTH, Stockholm

64. Goerigk L, Grimme S (2011) Efficient and Accurate Double-hybrid-meta-GGA density functionals—Evaluation with the extended GMTKN30 database for general main group thermochemistry, kinetics, and noncovalent interactions. J Chem Theory Comput 7:291–309. doi:10.1021/ct100466k

65. Benighaus T, DiStasio RA, Lochan RC et al (2008) Semiempirical double-hybrid density functional with improved description of long-range correlation. J Phys Chem A 112:2702–2712. doi:10.1021/jp710439w

66. Kozuch S, Gruzman D, Martin JML (2010) DSD-BLYP: A general purpose double hybrid density functional including spin component scaling and dispersion correction. J Phys Chem C 114:20801–20808. doi:10.1021/jp1070852

67. Riley KE, Pitoňák M, Jurečka P, Hobza P (2010) Stabilization and structure calculations for noncovalent interactions in extended molecular systems based on wave function and density functional theories. Chem Rev 110:5023–5063. doi:10.1021/cr1000173

68. Schwabe T, Grimme S (2007) Double-hybrid density functionals with long-range dispersion corrections: higher accuracy and extended applicability. Phys Chem Chem Phys 9:3397–3406. doi:10.1039/b704725h

69. Wu Q, Yang WT (2002) Empirical correction to density functional theory for van der Waals interactions. J Chem Phys 116:515–524. doi:10.1063/1.1424928

70. Grimme S (2006) Semiempirical GGA-type density functional constructed with a long-range dispersion correction. J Comput Chem 27:1787–1799. doi:10.1002/jcc.20495

71. Grimme S, Antony J, Ehrlich S, Krieg H (2010) A consistent and accurate ab initio parametrization of density functional dispersion correction (DFT-D) for the 94 elements H-Pu. J Chem Phys 132:154104. doi:10.1063/1.3382344

72. Klimeš J, Michaelides A (2012) Perspective: Advances and challenges in treating van der Waals dispersion forces in density functional theory. J Chem Phys 137:120901. doi:10.1063/1.4754130

73. Grimme S (2006) Seemingly simple stereoelectronic effects in alkane isomers and the implications for Kohn-Sham density functional theory. Angew Chem–Int Edit 45:4460–4464. doi:10.1002/anie.200600448

74. Cohen AJ, Handy NC (2001) Dynamic correlation. Mol Phys 99:607–615. doi:10.1080/00268970010023433

75. Levy M, March NH, Handy NC (1996) On the adiabatic connection method, and scaling of electron–electron interactions in the Thomas–Fermi limit. J Chem Phys 104:1989–1992. doi:10.1063/1.470954

76. Frisch MJ, et al. (2003) Gaussian 03, revision A. 1. Gaussian, Inc, Pittsburgh

77. Zhao Y, Truhlar DG (2006) A new local density functional for main-group thermochemistry, transition metal bonding, thermochemical kinetics, and noncovalent interactions. J Chem Phys 125:194101. doi:10.1063/1.2370993

78. Zhao Y, Truhlar DG (2008) The M06 suite of density functionals for main group thermochemistry, thermochemical kinetics, noncovalent interactions, excited states, and transition elements: Two new functionals and systematic testing of four M06-class functionals and 12 other functionals. Theor Chem Acc 120:215–241. doi:10.1007/s00214-007-0310-x

79. Zhang IY, Xu X (2012) A new generation density functional XYG3. Prog Chem 24:1023–1037

80. Slater JC (1960) Quantum theory of atomic structure, vol 2. McGraw-Hill, New York

81. Vosko SH, Wilk L, Nusair M (1980) Accurate spin-dependent electron liquid correlation endergies for local spin-density calculations–a critical analysis. Can J Phys 58:1200–1211. doi:10.1139/p80-159

82. Gorling A, Levy M (1994) Exact Kohn-Sham scheme based on perturbation theory. Phys Rev A 50:196–204. doi:10.1103/PhysRevA.50.196

83. Casida ME (1995) Generalization of the optimized-effective-potential model to include electron correlation: A variational derivation of the Sham-Schlüter equation for the exact exchange-correlation potential. Phys Rev A 51:2005–2013. doi:10.1103/PhysRevA.51.2005
84. Ivanov S, Bartlett RJ (2001) An exact second-order expression for the density functional theory correlation potential for molecules. J Chem Phys 114:1952–1955. doi:10.1063/1.1342809
85. Ivanov S, Levy M (2002) Accurate correlation potentials from integral formulation of density functional perturbation theory. J Chem Phys 116:6924–6929. doi:10.1063/1.1453952
86. Mori-Sánchez P, Wu Q, Yang WT (2005) Orbital-dependent correlation energy in density-functional theory based on a second-order perturbation approach: Success and failure. J Chem Phys 123:062204. doi:10.1063/1.1904584

Chapter 3
Benchmarking the Performance of DHDFs for the Main Group Chemistry

Abstract On one hand, chemistry is very rich. On the other hand, density functionals are all approximate and mostly contain empirical parameters, such that not every functional is equally applicable to every chemical problem. This has made benchmarking of the functional performance inevitable. Our focus here is to examine the performance of some fifth rung functionals, while selected results of the lower rung functionals are presented for comparison. We have examined the DHDFs' performance in the prediction of heats of formation (HOFs, Sect. 3.1), ionization potentials (IPs, Sect. 3.2), electron affinities (EAs, Sect. 3.2), bond dissociation energies (BDEs, Sect. 3.3), reaction barrier heights (RBHs, Sect. 3.4), and noncovalent interactions (NCIs, Sect. 3.5) using some well-established benchmarking data sets.

Keywords Heats of formation · Ionization potential · Electron affinity · Bond dissociation energy · Reaction barrier height · Noncovalent interaction

3.1 Heats of Formation Against the G3 Set

Heats of formation, ΔH_f^0, are among the most important chemical data, with which energy associated with a chemical reaction can be assessed. For stable molecules, experimental ΔH_f^0 may be found from some handbooks and/or from some websites.[e.g., 1–4] They are typically obtained from calorimetric measurements [5] using chemicals of very high purity. However, data for many other molecules still remain missing. This is especially true for reactive intermediates such as free radicals. Furthermore, the experimental data are frequently subject to substantial uncertainties [6]. Hence accurate computational chemistry methods are highly desirable such that a reliable prediction of thermochemical data can be made possible.

I. Y. Zhang and X. Xu, *A New-Generation Density Functional*,
SpringerBriefs in Molecular Science, DOI: 10.1007/978-3-642-40421-4_3,
© The Author(s) 2014

3.1.1 Performances of Various Rungs of the DFT Methods

The Gn paradigm was originally developed by Pople and co-workers for extrapolating levels of wavefunction based methods, as well as adjusting the empirical constants in the Gn methods to achieve increasingly accurate thermochemistry [7–9]. It has become a valuable dataset for developing density functionals to describe covalent bonding in the main group molecules. In particular, we use the G3 set of 223 molecules collected in 1999 (the G3/99 set) [9].

Table 3.1 summarizes the statistic data for the predicted ΔH_f^0 of the G3/99 set from various rungs of the DFT methods [10]. The calculations are based on the theoretical atomization enthalpy of a molecule corrected by the experimental atomization enthalpies of the constituent elements in their standard states at 298 K [9].

The mean absolute deviation (MAD) for the lowest rung LDA (e.g., SVWN5 [11, 12]) is 120.83 kcal/mol. This implies that LDA is not useful for thermochemistry.

GGAs on the second rung greatly reduce the errors. BPW91 [13, 14] leads to MAD of 8.78 kcal/mol, being one of the best GGAs up-to-date for molecules. The PBE functional [15] is less satisfactory for thermochemistry of molecules with MAD of 22.76 kcal/mol. There exists a large tendency of overbinding with maximum positive deviation (Max+) of 80.28 kcal/mol occurring at azulene.

The meta-GGA functionals on the third rung display further improvement. TPSS [16] gives MAD = 6.36 kcal/mol, being substantially better than its precedent PBE, while M06-L [17] and VSXC [18] lead to MAD of 5.82 and 3.51 kcal/mol, respectively.

Table 3.1 demonstrates that the hybrid functionals give an overall improvement for thermochemistry as compared to either pure GGAs or meta-GGAs. Thus the performance is significantly improved on going from pure BPW91 to three-parameter hybrid B3PW91 [13, 14, 19] (MAD = 3.85 kcal/mol). The popular functional, B3LYP [13, 19–21], is actually inferior (MAD = 4.74 kcal/mol). The best performer in this rung is ωB97X-D [22], whose MAD is 2.40 kcal/mol.

A recent important development in DFT is the M06 family of functionals (M06, and M06-2X) [23]. For the G3/99 set, these methods lead to MAD of 4.17 for M06, and 2.93 for M06-2X.

XYG3 [24] with the 6-311 + G(3df,2p) basis set [25–27] leads to MAD of 1.81 kcal/mol, being substantially better than the lower rung functionals (Table 3.1). While MC3BB [28] gives MAD = 3.81 kcal/mol, B2PLYP [29, 30] yields MAD of 2.74 kcal/mol. The latter was achieved by using a very large CQZV3P basis set including core-polarization [31], which was the way that B2PLYP was optimized [29].

We recall that B2PLYP employs its DFT portion for the SCF calculation to generate the orbitals from which the PT2 correction is computed. This is much like MP2 which uses HF for the SCF calculation [32]. Using just the DFT portion of B2PLYP with 6-311 + G(3df, 2p) leads to MAD = 174.20 kcal/mol for the G3/99 set, while HF gives MAD of 211.48 kcal/mol. The latter is ∼ 40 kcal/mol

Table 3.1 Theoretical errors [a] for heats of formation [b] (HOFs, kcal/mol) at 298 K for the G3/99 set [c]

		MAD [d]				Max + [e]	Max− [f]
		G2-1	G2-2	G3-3	G3		
1st Rung	SVWN	39.36	120.02	213.09	131.40	378.82 (azulene)	−0.30 (Li_2)
	SVWN5	36.03	110.76	195.51	120.83	345.45 (azulene)	−0.47 (Li_2)
	SPL	38.46	115.85	204.12	126.45	360.68 (azulene)	−0.16 (Li_2)
2nd Rung	BLYP	4.99	8.76	13.75	9.51	28.44 (NO_2)	−41.77 (n-octane)
	BPW91	5.18	9.31	10.82	8.78	31.89 (NO_2)	−24.10 ($Si(CH_3)_4$)
	PBE	8.37	22.62	33.48	22.76	80.28 (azulene)	−9.90 (Si_2H_6)
	BP86	9.82	25.14	36.99	25.28	71.10 (azulene)	−8.18 (SiF_4)
3rd Rung	M06-L	3.72	5.72	7.50	5.82	27.13 (C_2Cl_4)	−14.75 (PF_5)
	TPSS	4.70	7.46	6.22	6.36	25.04 (ClF_3)	−13.64 (SiF_4)
	VSXC	2.24	3.04	5.05	3.51	10.24(CS_2)	−12.75 (n-octane)
4th Rung	BHHLYP	2.16	3.46	8.20	4.74	8.03 (BeH)	−19.22 (SF_6)
	B3PW91	2.53	3.74	4.98	3.85	15.21(naphthalene)	−23.87 (SiF_4)
	B3LYP	2.16	3.46	8.20	4.74	8.03 (BeH)	−19.22 (SF_6)
	PBE0	2.85	6.43	10.48	6.91	35.69 (naphthalene)	−19.89 (SiF_4)
	BMK	2.24	3.29	3.30	3.03	12.52 (pyrimidine)	−13.32 (O_3)
	TPSSh	4.24	4.13	3.48	3.94	17.10 (Si_2H_6)	−23.90 (SiF_4)
	ωB97X	2.27	2.69	2.29	2.45	13.82 (C_2F_4)	−8.32 (Si_2)
	ωB97X-D	2.19	2.59	2.33	2.40	12.41 (C_2F_4)	−12.27 (SiF_4)
	M06-2X	1.90	3.20	3.36	2.93	17.39 (P_4)	−20.77 (O_3)
	M06	2.92	4.46	4.74	4.17	25.89 (C_2F_6)	−11.25 (O_3)
5th Rung	XYG3	1.53	1.78	2.06	1.81	6.28 (BCl_3)	−16.67 (SF_6)
	MC3BB	2.28	3.81	4.94	3.81	18.92 (naphthalene)	−10.38 (CN)
	B2PLYP [g]	1.36	2.00	4.68	2.74	6.60 (BeH)	−13.60 ($Si(CH_3)_4$)
	B2PLYP	1.85	3.67	7.84	4.63	8.01 (C_2F_4)	−20.37 (n-octane)
	B2PLYP-D	1.71	2.83	4.80	3.22	8.74 (C_2F_4)	−13.18 ($Si(CH_3)_4$)
	B2GP-PLYP	2.62	4.46	8.69	5.43	6.14 (C_2F_4)	−20.32 (n-octane)
Ab Initio	HF	74.61	191.57	336.54	211.48	0.46 (BeH)	−582.72 (n-octane)
	UMP2	7.34	11.12	13.33	10.93	48.34 (C_2F_6)	−29.21 ($Si(CH_3)_4$)
	QCISD(T) [h]	6.09	13.45	24.09	15.22	1.44 (Na_2)	−42.78(n-octane)
	G2 [i]	1.23	1.76	2.52	1.88	9.39 (C_2F_6)	−7.14 (SiF_4)
	G3 [i]	0.96	0.91	1.28	1.05	4.95 (C_2F_4)	−7.07 (PF_5)

[a] Errors, (kcal/mol, Expt. – Theo.). The geometries were optimized using B3LYP with the 6-311 + G(d, p) basis set. Analytical vibrational frequencies were calculated at the same level and scaled by 0.9877 to estimate zero-point energies. Single point calculations are performed with the 6-311 + G(3df, 2p) basis set [25–27].
[b] Experimental data are from Ref. [9].
[c] The G3/99 set are usually divided into three subsets, G2-1, G2-2 and G3-3, of increased molecular size.
[d] Mean absolute deviations.
[e] Maximum positive deviations.
[f] Maximum negative deviations.
[g] Data from Ref. [30].
[h] Data from Ref. [40].
[i] Calculated with the Gn theory [7–9]

worse. The complete B2PLYP method leads to MAD = 4.63 kcal/mol, while the corresponding MAD associated with MP2 is 10.93 kcal/mol. Due to the approximation adopted by restricting the minimization in Eq. 2.31 to single-determinant wavefunction Φ [32], density in B2PLYP does not corresponds to the ground state density by construction just as in the HF method. While B2PLYP density should be an improvement over HF density, due to the partial correlation embedded in the complement HK functional $\overline{E}_{Hxc}^{\lambda}[\rho]$ (see Sect. 2.3). B2PLYP is certainly an improvement over MP2 in terms of the G3 set for covalent bonds.

XYG3 is rooted within the adiabatic connection formulism [33–37] and the Görling-Levy theory [38] of coupling-constant perturbation expansion, we consider it very important to have accurate KS orbitals to provide an accurate density and the zero-order approximation for perturbation theory [24]. It has been shown that B3LYP densities are similar to those from CCSD(T) ab initio wavefunctions (for the molecules discussed in Ref. [39]). Hence, XYG3 adopts B3LYP densities, as well as the large energy terms $T[\rho]$, $J[\rho]$, and $V_{ext}[\rho]$, but only updates the small energy term $E_{xc}[\rho]$.

From Table 3.1 it is clear that there is a general tendency for the improvement of the DFA performance along the rungs upwards. MADs are gradually reduced from 120.83 kcal/mol for SVWN5 [11, 12] to 8.78 for BPW91 [13, 14], to 3.51 for VSXC [18], to 2.40 for ωB97X-D [22], and to 1.81 for XYG3 [24].

The G3 method leads to a MAD of only 1.05 kcal/mol, while that given by the G2 method is 1.88 kcal/mol [8, 9]. Gn is a composite method, based on the 6-311G** or 6-31G(d) basis sets but with several basis set extensions [7–9, 25–27]. Electron correlation is treated by the MP perturbation theory and by quadratic configuration interaction (QCISD(T)). Even though we have grouped the Gn method under the title of ab initio, we must not overlook the fact that it includes an empirical 'high-level correction (HLC)'. Removing this HLC leads to much poorer thermochemistry as shown by the QCISD(T) results listed in Table 3.1 (MAD = 15.22 kcal/mol) [40]. Thus the current generation of DFT functionals lead to HOFs significantly better than the standard ab initio methods (e.g., QCISD(T) with basis sets of triple-zeta quality). There is no doubt, however, that increasing the basis set size will significantly improve the QCISD(T) performance. Coupled-cluster based method at sufficiently large basis set, although much expensive, has set up the gold standard that DFAs are trying to approach to.

3.1.2 Basis Set Dependence

It has to be noted that the good behavior of XYG3 for HOFs as listed in Table 3.1 is partly because that XYG3 was fitted against the G3/99 set with the 6-311 + G(3df, 2p) basis set. HOFs was found to be subject to a large basis set dependence [10, 41], since all chemical bonds are broken during the atomization process. The basis set dependence of XYG3, along with those of B3LYP and MP2, have been investigated [10, 41]. The basis sets examined included [25–27] B1: 6-311 + G(d, p), B2: 6-311 + G(2d, p), B3: 6-311 + G(2d, 2p), B4: 6-311 + G(3d, 2p), B5: 6-311 + G(2df, p), B6: 6-311 + G(2df, 2p), B7: 6-311 + G(3df), B8: 6-311 + G(3df, p), B9: 6-311 + G(3df, 2p), and B10: 6-311 ++G(3df, 3pd), where B9 is the

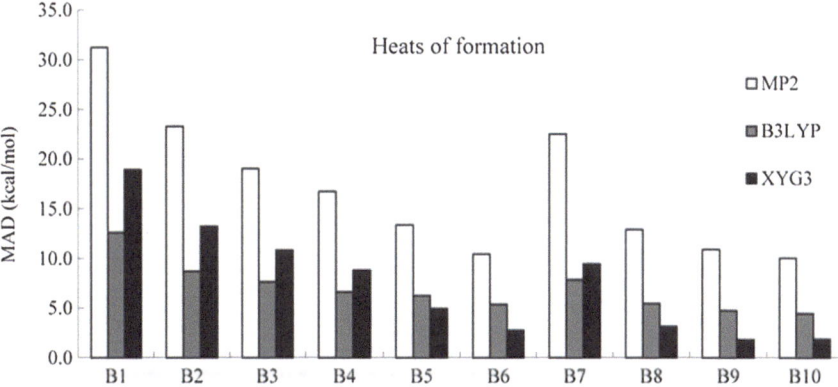

Fig. 3.1 Mean absolute deviations (MADs): basis set dependence for heats of formation against the G3/99 set. B1: 6-311 + G(d, p), B2: 6-311 + G(2d, p), B3: 6-311 + G(2d, 2p), B4: 6-311 + G(3d, 2p), B5: 6-311 + G(2df, p), B6: 6-311 + G(2df, 2p), B7: 6-311 + G(3df), B8: 6-311 + G(3df, p), B9: 6-311 + G(3df, 2p), and B10: 6-311 ++G(3df, 3pd)

designed basis set. This choice echoes the common wisdom in the molecular orbital (MO) theory that a triple-zeta basis set is relatively complete for moderate accuracy and the major source of errors in calculating chemical reaction energies such as HOFs comes from the incompleteness of the polarization functions [42, 43].

Figure 3.1 depicts the MADs for HOFs against the G3/99 set with this set of basis sets [10, 41]. B3LYP shows a mild basis set dependence. MAD associated with B3LYP/B1 is 12.63 kcal/mol, which decreases to 4.74 with the B10 basis set. This is expected as adding more polarization functions improves the description of the molecules, reducing B3LYP's tendency of underestimating the stability of the molecules. Figure 3.1 clearly shows that MP2 is more basis set dependent than B3LYP. MAD associated with MP2 spans a range of 13.30 kcal/mol from B1 (MAD = 31.35) to B10 (17.95 kcal/mol), as opposed to the B3LYP range of 7.89 kcal/mol. XYG3 has inherited the strong basis set sensitivity of MP2. MAD for XYG3/B1 is as high as 18.93 kcal/mol. When augmented with suitable number of polarization functions, XYG3 starts to behave significantly better than B3LYP. Overall, XYG3 presents an improvement over both B3LYP and MP2 in HOF predictions with B5, B6, B8, B9 and B10 basis sets [41].

3.1.3 Molecular Size Dependence of Representative DHDFs

The G3/99 set is usually divided into three subsets, G2-1, G2-2, and G3-3 [7–9]. While the G2-1 set consists of 55 molecules with the maximum number of non-hydrogen atoms of 3, the G2-2 set and the G3-3 set are made of 93 and 75 molecules up to 6 and 10 nonhydrogen atoms, respectively. The averaged non-hydrogen atoms are 1.6 (G2-1), 3.6 (G2-2), and 5.8 (G3-3), indicating an increased size from G2-1 to G2-2 and G3-3.

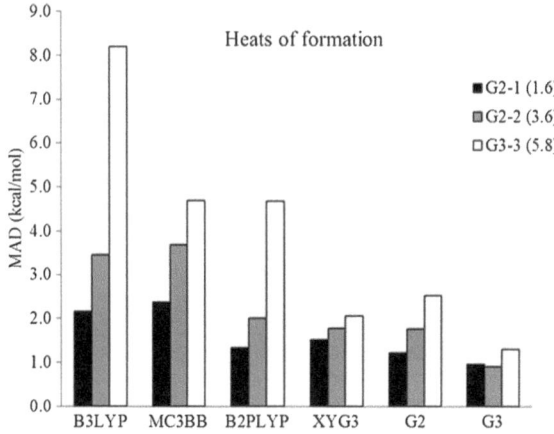

Fig. 3.2 Mean absolute deviations (MADs): molecular size dependence for heats of formation against the G3/99 set

Figure 3.2 depicts the molecular size dependence against the G3/99 set for the methods we are interested in. More data are presented in Table 3.1 [10]. B3LYP leads to errors that increase dramatically with size [42, 43], with MAD = 2.12 kcal/mol (G2-1), 3.69 (G2-2), and 8.97 (G3-3). B2PLYP (at the 6-311 + G(3df, 2p) level) does not improve over B3LYP, leading to MADs of 1.85 (G2-1), 3.70 (G2-2), and 7.83 kcal/mol (G3-3). Size dependence is mild for MC3BB [28], giving MADs of 2.28 (G2-1), 3.81 (G2-2), and 4.97 kcal/mol (G3-3). For XYG3 [24], we obtain MADs of 1.52 (G2-1), 1.79 (G2-2), and 2.06 kcal/mol (G3-3), which exhibits the best description for larger molecules [10, 24].

3.2 Ionization Potentials, and Electron Affinities Against the G2-1 Set

Ionization potential (IP) and electron affinity (EA) are essential molecular properties [44–46]. They define many useful concepts such as electronegativity [47], chemical potential [44], hardness and softness [45], as well as electrophilicity and nucleophilicity [46], etc. They can be used in assessing the electron donating and accepting abilities of a system involved in any redox processes [48–50]. IP and EA have been widely employed in understanding the electron transfer processes occurring in gas phase or in condensed phase, and are of fundamental importance in setting up structure–reactivity relationships to aid the design of new materials [51–54].

3.2.1 Error Statistics for Calculated Ionization Potentials

Table 3.2 lists the error statistics for IPs of 38 molecules in the G2-1 set [8, 10, 55]. Generally, charged species are more inhomogeneous than the corresponding neutral systems. Thus it is expected that LDA leads to the worst results for IP

Table 3.2 Statistic theoretical errors[a,b,c] for calculated ionization potentials (IPs, eV) at 0 K for the G2-1 Set (38 systems)

		AD	MAD	RMS	Max +	Max-
1st Rung	SVWN	−0.663	0.663	0.698	NA[g]	−1.15 (F → F$^+$)
	SVWN5	−0.175	0.224	0.269	0.30 (Be → Be$^+$)	−0.56 (O$_2$ → O$_2^+$)
	SPL	−0.190	0.232	0.279	0.58 (H → H$^+$)	−0.73 (O$_2$ → O$_2^+$)
2nd Rung	BLYP	0.078	0.200	0.240	0.43 (Cl$_2$ → Cl$_2^+$)	−0.56 (O → O$^+$)
	BPW91	0.064	0.229	0.377	1.71 (H$_2$S → H$_2$S$^+$ ^{2}A$_1$)	−0.46 (O → O$^+$)
	PBE	−0.004	0.161	0.200	0.34 (Cl$_2$ → Cl$_2^+$)	−0.46 (O → O$^+$)
	BP86	−0.040	0.231	0.367	1.60 (H$_2$S → H$_2$S$^+$ ^{2}A$_1$)	−0.63 (O → O$^+$)
3rd Rung	M06-L	0.077	0.193	0.239	0.64 (Na → Na$^+$)	−0.59 (O$_2$ → O$_2^+$)
	TPSS	0.037	0.173	0.205	0.31 (C$_2$H$_2$ → C$_2$H$_2^+$)	−0.45 (B → B$^+$)
	VSXC	0.046	0.192	0.346	1.64 (H$_2$S → H$_2$S$^+$ ^{2}A$_1$)	−0.33 (N → N$^+$)
4th Rung	BHHLYP	0.076	0.213	0.281	0.47 (C$_2$H$_4$ → C$_2$H$_4^+$)	−0.99 (O$_2$ → O$_2^+$)
	B3PW91	−0.056	0.159	0.208	0.32 (Be → Be$^+$)	−0.72 (O$_2$ → O$_2^+$)
	B3LYP	−0.087	0.162	0.226	0.20 (Be → Be$^+$)	−0.79 (O$_2$ → O$_2^+$)
	PBE0	−0.001	0.165	0.204	0.34 (Be → Be$^+$)	−0.68 (O$_2$ → O$_2^+$)
	BMK	−0.034	0.161	0.236	0.52 (Be → Be$^+$)	−0.81 (O$_2$ → O$_2^+$)
	TPSSh	0.074	0.214	0.347	1.73 (H$_2$S → H$_2$S$^+$ ^{2}A$_1$)	−0.53 (O$_2$ → O$_2^+$)
	ωB97X	0.007	0.135	0.187	0.46 (Be → Be$^+$)	−0.63 (O$_2$ → O$_2^+$)
	ωB97X-D	−0.006	0.132	0.187	0.49 (Be → Be$^+$)	−0.67 (O$_2$ → O$_2^+$)
	M06-2X	−0.011	0.119	0.196	0.28 (SiH$_4$ → SiH$_4^+$)	−0.82 (O$_2$ → O$_2^+$)
	M06	0.027	0.159	0.211	0.39 (Be → Be$^+$)	−0.76 (O$_2$ → O$_2^+$)
5th Rung	XYG3	0.010	0.057	0.075	0.20 (N$_2$ → N$_2^+$,$^2\Sigma_g$)	−0.16 (O$_2$ → O$_2^+$)
	MC3BB	0.070	0.120	0.150	0.42 (Be → Be$^+$)	−0.40 (O$_2$ → O$_2^+$)
	B2PLYP	0.049	0.109	0.130	0.31 (Be → Be$^+$)	−0.31 (O$_2$ → O$_2^+$)
	B2PLYP-D	0.054	0.110	0.131	0.31 (Be → Be$^+$)	−0.31 (O$_2$ → O$_2^+$)
	B2GP-PLYP	0.050	0.101	0.123	0.33 (Be → Be$^+$)	−0.31 (O$_2$ → O$_2^+$)
Ab initio	HF	0.954	1.005	1.135	1.82 (Be → Be$^+$)	−0.84 (O$_2$ → O$_2^+$)
	UMP2	0.077	0.163	0.218	0.50 (Be → Be$^+$)	−0.69 (CS → CS$^+$)
	MP4SDQ	0.116	0.150	0.173	0.33 (S → S$^+$)	−0.38 (CS → CS$^+$)
	QCISD(T)	0.106	0.111	0.125	0.27 (S → S$^+$)	0.11 (CS → CS$^+$)

[a] IPs are calculated as ground state energy differences between the neutral species and the corresponding ionic species [8, 10, 55]. As in the G2 method [8], the geometries were optimized using MP2(full) with the 6-31G(d) basis set. Analytical vibrational frequencies were calculated at the level of HF/6-31G(d) and scaled by 0.8929 to estimate zero-point energies. Single point DFT calculations were performed with the 6-311 + G(3df, 2p) basis set [25–27].
[b] Experimental data are from Refs. [8].
[c] AD: Averaged deviations (Expt. – Calc.), MAD: Mean absolute deviations, RMS: Root-mean-square errors, Max + : Maximum positive deviations, Max-: Maximum negative deviations

calculations. This is indeed true for SVWN [11, 12] (MAD = 0.663 eV), but not necessary so for SVWN5 [11, 12] and SPL [11, 56] (MADs = 0.224 and 0.232 eV, respectively). This suggests that the correlation functionals play an important role.

Starting from that of SVWN, one may see that the second rung functionals of GGAs dramatically improve the prediction of IPs over LDA(SVWN). Actually,

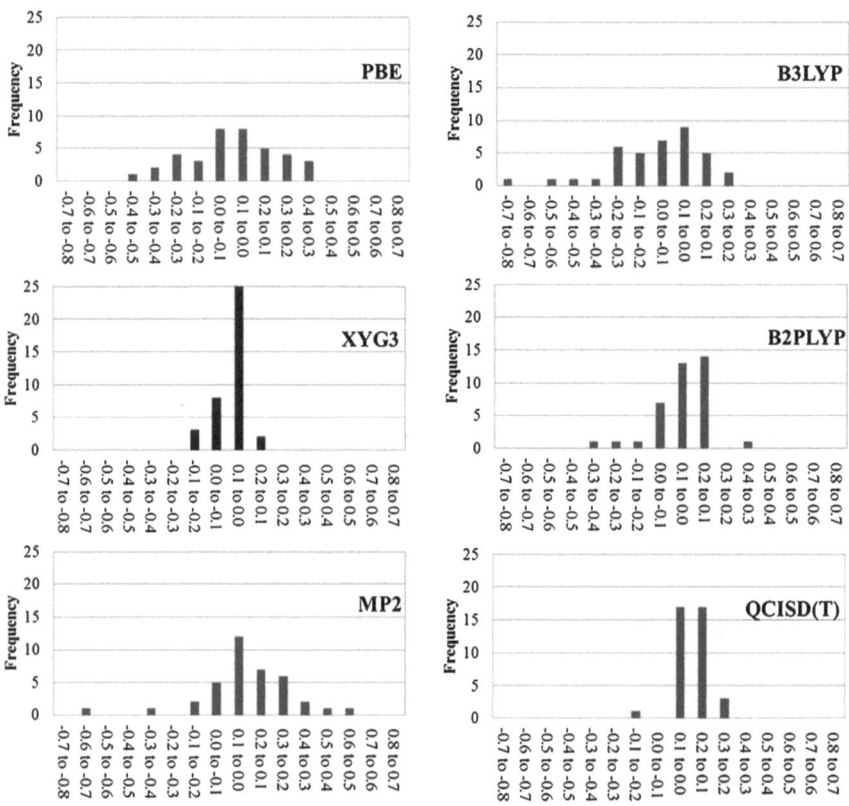

Fig. 3.3 Histogram of deviations (Expt. – Calc.) for 38 ionization potentials (IPs) in the G2-1 set

GGAs on average give an MAD of 0.20 eV, being close to that of SVWN5, while PBE [15] gives an MAD of 0.161 eV, which is the best GGA for IP predictions. From Table 3.2 it can be seen that meta-GGAs slightly outperform GGAs, although TPSS [16] (MAD = 0.173 eV) is slightly worse than its precedent PBE.

As ionization may create or quench a singly occupied orbital, the IP error may therefore be related to the self-interaction error [57] which differs on the amount between the neutral and the corresponding charged species. Hence, one sees, from Table 3.2, a steady improvement for IP prediction along the ladder up to hybrid functionals, possibly due to the mitigation of the self-interaction error because of the introducing of the HF exchange. The majority of the 4th rung functionals give an MAD around 0.16 eV, while some hybrid functionals, (e.g., ωB97X and ωB97X-D [22]) give an MAD around 0.13 eV. M06-2X [23] is outstanding, giving an MAD of only 0.119 eV for the G2-1 set.

As it should be expected, the 5th rung functionals should further improve the accuracy because of the further improvement on the correlation part. Indeed, all DHDFs give MADs below 0.12 eV. Significantly, XYG3 gives an MAD of only 0.057 eV for the G2-1 set [8, 10, 55].

It is worthwhile to make a comparison to the IP performances of the wave-function based methods. Clearly, HF (MAD = 1.005 eV) is useless for the IP prediction, emphasizing again the importance of correlation effects. MP2 has dramatically reduced MAD to 0.163 eV, which has been further reduced to 0.150 eV for MP4SDQ and to 0.111 eV for QCISD(T). It should be emphasized that basis set requirement is generally higher for wavefunction based methods. With larger basis set than 6-311 + G(3df, 2p) used here, we expect that QCISD(T) will lead to more satisfactory results, albeit at higher expense.

Figure 3.3 shows the histogram of deviations (Expt. – Calc.) for selected methods in the predictions of 38 IPs in the G2-1 set [55]. PBE errors scatter around a range with more negative deviations, indicating that neutral species are sub-stantially overbound relative to the ions. B3LYP behaves worse than PBE. Both B2PLYP and XYG3 are outstanding for IP calculations. In particular, 25 out of 38 entries are within the 0.0–0.1 eV error range for XYG3, showing the predictive power of this functional. We emphasize that ions were not included in the training set for XYG3. The three mixing parameters in XYG3 were optimized by using only heats of formation of the G3 set where all species are neutral.

3.2.2 Error Statistics for Calculated Electron Affinities

Table 3.3 lists the statistics for EA calculations of the total 25 systems in the G2-1 set [8, 10, 55]. The DFA performances may be compared with those of wave-function methods such as MP2 and QCISD(T).

Similar to the IP calculations, SVWN [11, 12] also performs quite differently from SVWN5 [11, 12] and SPL [11, 56] for EA calculations. SVWN gives an MAD = 0.750 eV, significantly worse than MADs of 0.289 and 0.311 eV for SVWN5 and SPL, respectively. Unlike situations in IP calculations, Table 3.3 clearly demonstrates that GGAs and meta-GGAs perform considerably better than LDAs for EA calculations. The best GGA and meta-GGAs lead to MADs of 0.094 (BPW91 [13, 14]) and 0.104 eV (TPSS [16], VSXC [18]). Some hybrid functionals (e.g., ωB97X-D [22]) and DHDFs (e.g., XYG3 [24]) can still improve the accuracy to some extent, giving MADs around 0.08 eV. Indeed, the error histograms displayed in Fig. 3.4 suggest that XYG3 is the most satisfactory DFT for EA predictions.

From theoretical point of view, there has been debate in the literature, concerning whether conceptually DFT methods are suitable for calculating electron affinities [58–60]. It has been argued that two artifacts that combine fortunately in the right way for error cancelation. On one hand, the 'self-interaction error' causes the Kohn–Sham orbital energies to shift upwards artificially, leading to a positive (unstable) orbital energy for the highest occupied orbital of the anion. On the other hand, an artificial stabilization is provided by employing a finite basis set with functions localized at the anion. This debate continues [57].

Table 3.3 Statistic theoretical errors [a,b,c] for electron affinities (EAs, eV) at 0 K for the G2-1 Set (25 systems)

		AD	MAD	RMS	Max +	Max-
1st Rung	SVWN	−0.750	0.750	0.766	NA[g]	−1.15 (F ← F⁻)
	SVWN5	−0.289	0.289	0.269	NA[g]	−0.63 (F ← F⁻)
	SPL	−0.311	0.311	0.345	NA[g]	−0.67 (F ← F⁻)
2nd Rung	BLYP	−0.026	0.105	0.135	0.19 (Si ← Si⁻)	−0.36 (Cl_2 ← Cl_2^-)
	BPW91	−0.060	0.094	0.123	0.12 (S_2 ← S_2^-)	−0.27 (C ← C⁻)
	PBE	−0.086	0.102	0.133	0.10 (S_2 ← S_2^-)	−0.29 (C ← C⁻)
	BP86	−0.211	0.211	0.232	NA[g]	−0.39 (Cl_2 ← Cl_2^-)
3rd Rung	M06-L	0.118	0.160	0.186	0.37 (OH ← OH⁻)	−0.28 (Cl_2 ← Cl_2^-)
	TPSS	0.015	0.104	0.122	0.19 (OH ← OH⁻)	−0.27 (Cl_2 ← Cl_2^-)
	VSXC	−0.017	0.104	0.148	0.25 (O_2 ← O_2^-)	−0.54 (Cl_2 ← Cl_2^-)
4th Rung	BHHLYP	0.198	0.248	0.285	0.58 (OH ← OH⁻)	−0.29 (Cl_2 ← Cl_2^-)
	B3PW91	−0.030	0.103	0.126	0.17 (OH ← OH⁻)	−0.29 (Cl_2 ← Cl_2^-)
	B3LYP	−0.084	0.106	0.144	0.06 (OH ← OH⁻)	−0.45 (Cl_2 ← Cl_2^-)
	PBE0	0.036	0.128	0.146	0.29 (OH ← OH⁻)	−0.20 (Cl_2 ← Cl_2^-)
	BMK	0.007	0.106	0.127	0.22 (F ← F⁻)	−0.32 (CN ← CN⁻)
	TPSSh	0.053	0.130	0.153	0.29 (OH ← OH⁻)	−0.26 (Cl_2 ← Cl_2^-)
	ωB97X	0.009	0.083	0.106	0.17 (Si ← Si⁻)	−0.25 (CN ← CN⁻)
	ωB97X-D	−0.013	0.079	0.100	0.15 (OH ← OH⁻)	−0.22 (Cl_2 ← Cl_2^-)
	M06-2X	0.051	0.103	0.126	0.25 (F ← F⁻)	−0.17 (CN ← CN⁻)
	M06	0.048	0.095	0.116	0.24 (SiH_2 ← SiH_2^-)	−0.23 (NO ← NO⁻)
5th Rung	XYG3	0.058	0.080	0.090	0.16 (CH_3 ← CH_3^-)	−0.18 (Cl_2 ← Cl_2^-)
	MC3BB	0.132	0.175	0.188	0.29 (NH ← NH⁻)	−0.26 (CN ← CN⁻)
	B2PLYP	0.056	0.090	0.102	0.17 (CH_3 ← CH_3^-)	−0.22 (Cl_2 ← Cl_2^-)
	B2PLYP-D	0.056	0.091	0.104	0.17 (CH_3 ← CH_3^-)	−0.23 (Cl_2 ← Cl_2^-)
	B2GP-PLYP	0.083	0.114	0.124	0.21 (CH_3 ← CH_3^-)	−0.19 (CN ← CN⁻)
Ab initio	HF	1.148	1.148	1.283	2.21 (F ← F⁻)	NA
	UMP2	0.079	0.166	0.224	0.37 (P ← P⁻)	−0.78 (CN ← CN⁻)
	MP4SDQ	0.175	0.208	0.229	0.36 (NH ← NH⁻)	−0.41 (CN ← CN⁻)
	QCISD(T)	0.135	0.135	0.148	0.23 (NH ← NH⁻)	NA

[a] EAs are calculated as ground state energy differences between the neutral species and the corresponding ionic species [8, 10, 55]. As in the G2 method [8], the geometries were optimized using MP2(full) with the 6-31G(d) basis set. Analytical vibrational frequencies were calculated at the level of HF/6-31G(d) and scaled by 0.8929 to estimate zero-point energies. Single point DFT calculations were performed with the 6-311 + G(3df, 2p) basis set [25–27]. [b] Experimental data are from Refs [8]. [c] AD: Averaged deviations (Expt. – Calc.), MAD: Mean absolute deviations, RMS: Root-mean-square errors, Max + : Maximum positive deviations, Max-: Maximum negative deviations, NA: Not applied

From practical point of view, it is encouraging to see from Table 3.3 that many DFT methods outperform MP2 (MAD = 0.166 eV) and QCISD(T) (MAD = 0.135 eV) for EA calculations. With larger basis set than 6-311 + G(3df, 2p) used here, there is no doubt that QCISD(T) will lead to more satisfactory results, although its steep scaling will prevent its applications to larger systems.

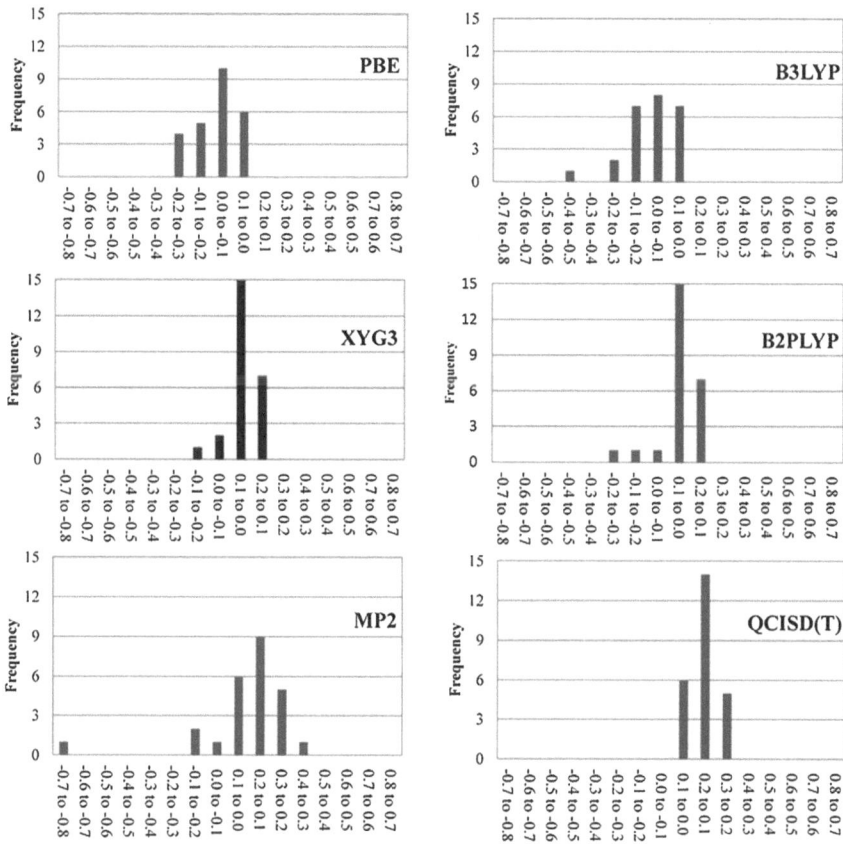

Fig. 3.4 Histogram of deviations (Expt. – Calc.) for 25 electron affinities (EAs) in the G2-1 set

3.3 Bond Dissociation Enthalpies Against the BDE/07 Set

Covalent bond dissociation enthalpy (BDE) is defined as energy absorbed when a bond is cleaved by homolysis. It is a fundamental concept in chemistry, being widely used in pursuit of the understanding of a diversity of chemical processes such as atmospheric and combustion reactions, or enzymatic catalysis, etc. Nevertheless, the majority of experimental BDE data suffers from an uncertainty of 1–2 kcal/mol [1–3, 6, 8], and some of them are contradictory to each other. Theory provides a valuable alternative. In particular, it offers a way to provide BDEs of unknown species or unstable species that are not amenable to any experimental techniques. Hence, developing accurate theoretical methods for BDE prediction is vitally important.

3.3.1 Performances of Various Rungs of the DFT Methods

We calculated BDE of a single bond according to the enthalpy change of the following reaction in the gas phase at 298 K and 1 atm:

$$X - Y(g) = X \cdot (g) + Y \cdot (g)$$

$$BDE = \Delta_r H_{298}^0 = \Delta_f H_{298}^0(X \cdot) + \Delta_f H_{298}^0(Y \cdot) - \Delta_f H_{298}^0(X - Y) \qquad (3.1)$$

where we supplied the experimental or calculated HOF with the given method for each species. When X or Y happened to be an atom, we used the experimental HOF [7–9].

Twenty-seven radicals and seventy-six molecules contained in the G3/99 set have been used to set up ninety-two bond dissociation reactions (the BDE92/07 set). Five radicals and thirty-nine molecules that go beyond the G3/99 set have further been included to set up fifty additional bond dissociation reactions, leading to the so-called BDE142/07 set. The bond types include C–H, X–H, C–C, C–O, C–N, C–F, C–Cl, C–S and X–Y (X, $Y \neq$ C, H). The lowest BE is that for CH$_3$–CO, 10.98 kcal/mol, while the highest BE is that for NC–CN, 136.50 kcal/mol. Heats of formation span from -321.3 (C$_2$F$_6$) to 135.1 (CCH) kcal/mol.

As HOFs are routinely calculated via atomization energies, which present a harsh chemistry where every bond in the molecule is broken, it is generally believed that errors associated with BDE calculations are smaller than HOF calculations. Contrary to this general belief, Table 3.4 and Fig. 3.5 show that a good prediction of HOFs does not necessarily guarantee a good performance for BDE prediction [61–63].

As indicated in Eq. 3.1, if errors in HOFs for radicals and the parent molecules are of the opposite sign, there will be an error accumulation in the prediction of BDE. This is obviously the case for B3LYP [13, 19–21] as shown in Fig. 3.5 for the BDE142/07 set where MAD for HOFs is 3.96 which increases to 6.14 kcal/mol

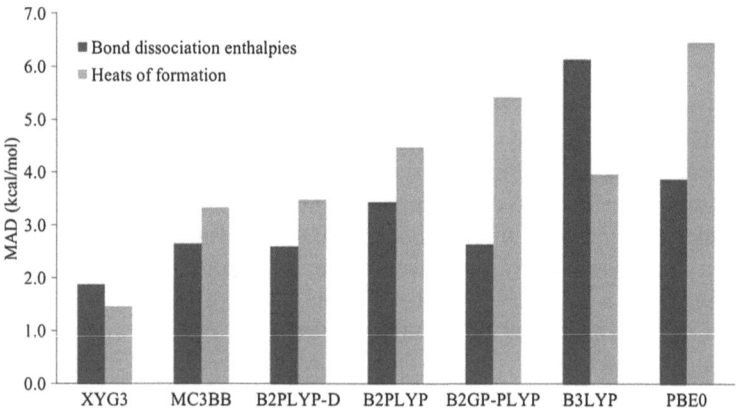

Fig. 3.5 Mean absolute deviation for the performance of some representative functionals in calculating heats of formation and bond dissociation enthalpies

Table 3.4 Statistic theoretical errors (in kcal/mol)[a,b] for bond dissociation enthalpies (BDEs) at 298 K (the BDE142/07 set) and the associated set of heats of formations (HOFs)

	Method	BDE				HOF			
		AD	MAD	Max+	Max-	AD	MAD	Max+	Max-
1st Rung	SPL	−16.79	16.81	1.02	−48.95	137.63	137.63	430.65	NA
	SVWN	−17.79	17.79	0.30	−49.70	142.37	142.37	452.75	NA
	SVWN5	−16.01	16.03	1.09	−46.80	130.71	130.71	412.44	NA
2nd Rung	BLYP	7.39	8.15	22.71	−13.22	−2.52	10.05	41.66	−43.41
	PBE	1.92	4.74	14.46	−21.40	24.21	24.48	89.89	−9.90
	BPW91	5.35	6.71	19.82	−16.44	6.23	10.25	49.23	−22.75
	BP86	3.13	5.15	16.78	−19.37	28.56	28.56	79.92	NA
3rd Rung	M06-L	3.45	4.56	12.01	−10.58	1.69	4.97	21.02	−9.53
	TPSS	5.96	6.78	19.05	−10.19	6.91	7.02	23.33	−3.99
	VSXC	4.86	5.36	12.18	−7.29	0.92	3.28	14.91	−8.81
4th Rung	BHHLYP	7.53	8.12	21.23	−9.01	−29.41	29.51	2.82	−90.79
	B3PW91	4.25	5.40	16.80	−12.76	2.79	3.85	18.48	−13.22
	B3LYP	5.74	6.14	18.55	−.78	−2.11	3.96	8.50	−19.81
	B3P86	1.41	2.20	13.01	−9.68	29.93	29.93	89.51	NA
	PBE0	2.91	3.87	12.46	−8.94	5.82	6.46	40.99	−7.24
	BMK	0.45	2.16	9.05	−11.07	1.25	2.18	6.99	−5.93
	TPSSh	6.18	6.61	18.17	−8.78	0.95	3.48	17.10	−8.68
	M06-2X	−0.67	2.06	5.16	−12.07	0.37	2.31	15.80	−9.54
	M06	0.57	2.32	7.12	−8.94	1.99	3.51	25.89	−4.24
5th Rung	XYG3	1.62	1.87	7.34	−3.35	0.01	1.45	6.87	−7.65
	MC3BB	−1.61	2.65	4.75	−21.46	1.93	3.33	22.36	−10.38
	B2PLYP	2.97	3.44	9.26	−6.61	−3.37	4.47	10.24	−21.16
	B2PLYP-D	2.14	2.60	6.25	−6.82	−2.70	3.48	8.94	−10.99
	B2GP-PLYP	1.85	2.64	6.33	−10.77	−5.08	5.42	3.90	−20.38
Ab initio	HF	32.04	32.27	80.66	−6.63	−231.61	231.61	−24.33	−666.43
	UMP2	−8.29	9.34	8.62	−57.10	−0.18	12.06	48.34	−25.88
	G2 [d]	−2.12	2.35	3.85	−8.88	0.44	1.76	9.39	−7.11
	G3 [d]	−0.16	1.35	4.62	−6.56	0.17	0.80	3.60	−4.85

[a] B3LYP/6–311 + G(d,p) optimized geometries are adopted for all calculations. ZPEs are scaled with 0.9877. B3LYP/ 6–311 + G(3df, 2p) is used for single point energy calculations [25–27]. Experimental data are from Refs. [8, 9].
[b] AD: Averaged deviations (Expt. – Calc.), MAD: Mean absolute deviations, RMS: Root-mean-square errors, Max + : Maximum positive deviations, Max-: Maximum negative deviations, NA: Not applied

for BDEs. On the other hand, even though PBE0 [15, 58, 64] leads to MAD of 6.46 kcal/mol for HOFs, the corresponding MAD for BDEs is reduced to 3.87 kcal/mol, benefiting from error cancelations between HOFs of molecules and radicals. As shown in Table 3.4, such favorable error cancelations are most significant for LDA(e.g., SPL [11, 55]), GGA(e.g., PBE [15]), and the HF method.

As for DHDFs, we find that MAD associated with XYG3 is 1.45 kcal/mol for HOFs, which is slightly increased to 1.87 kcal/mol for BDEs, being quite satisfactory for both quantities. MADs for HOFs are 2.98 (B2PLYP-D [30]), 3.33 (MC3BB [28]), 4.47 (B2PLYP [29]), and 5.42 (B2GP-PLYP [65]), the corresponding errors for BDEs are 2.65 (MC3BB), 2.51 (B2PLYP-D), 2.64 (B2GP-PLYP), and 3.44 (B2PLYP). In comparison, G2 and G3 give MADs of 1.76, and 0.80 kcal/mol, respectively, for HOFs, and 2.35, and 1.35 kcal/mol, respectively, for BDEs.

3.3.2 Error Statistics for Different Bond Types

Statistical analyses are also performed according to bond types in the BDE142/ 07 set (see Table 3.5). For 16 comparisons of C–H bonds, B3LYP leads to MAD = 2.70 kcal/mol, which is one of the best results, on average, as compared to its own performance against other bond types. XYG3 reduces B3LYP MAD to only 1.06 kcal/mol. M06-2X and B2PLYP-D give MADs of 1.49 and 1.98 kcal/mol, respectively, while all other DFAs lead to MADs higher than 2.0 kcal/mol. G3 is most satisfactory for C–H bonds, whose MAD is only 0.50 kcal/mol; while G2 is less satisfactory as compared to G3 with MAD being 1.53 kcal/mol.

For 12 comparisons of X–H bonds ($X \neq$ C), XYG3 performance is degraded, with MAD being 2.41 kcal/mol. The best DFA performer is M06, whose MAD is 1.38 kcal/mol, while MC3BB is the second best (MAD = 1.66 kcal/mol). Note that MAD is more than doubled for G3 to 1.35 kcal/mol, it is slightly improved to 1.42 kcal/mol for G2, as compared to their own performance for C–H bonds.

The C–C bond type presents the largest subset in our benchmarking set (44 comparisons). It covers various chemical situations where a C–C bond is embedded. MAD of B3LYP is significantly increased to 8.03 kcal/mol. The most difficult cases are when the carbon is highly alkylated [66–68]. Both XYG3 (MAD = 1.69 kcal/mol) and M06-2X (MAD = 1.41 kcal/mol) present a significant improvement, which are comparable to the G3 method (MAD = 1.23 kcal/ mol), while G2 is inferior, giving MAD of 2.51 kcal/mol. This suggests that both XYG3 and M06-2X have taken good care of middle range correlations [68–70]. In fact, long-range dispersive correlations are also believed to be taken into account to some extent [24].

We have also included fifteen X–Y bonds, where X, and Y are neither C nor H atoms [63]. Accurate description of these bonds is a great challenge as it may involve subtle balance between hyperconjugative effect and steric repulsion of lone pair electrons as in HO–OH and $H_2N–NH_2$. G3 gives MAD of 2.11 kcal/mol, being one of the worst sets of its own according to the bond types. MC3BB, being the best method for this bond type, leads to MAD of 1.45 kcal/mol, whereas MADs of other DHDFs are within 3.0–3.6 kcal/mol.

B3LYP performance for C–X (X = O, N, S) in alcohols, ethers, thiols, sulfides, and amines is poor. For a total of 40 comparisons in three subsets, MAD is as high as 7.4 kcal/mol. Increasing alkylation again leads to increasing errors. PBE0 is better, which nearly halves the errors. The M06 family is even better. While M06 is good for C–O/C–N bonds, with MADs of 1.67/1.79 kcal/mol; M06-2X is good for C–S with MAD of 1.06 kcal/mol. XYG3 is fair, with MADs around 2 kcal/mol. The best DHDF is MC3BB, whose MADs are 1.32, 1.98, and 1.04 kcal/mol for C–X (X = O, N, S, respectively). Even the Gn methods are not

Table 3.5 Bond dissociation enthalpies (BDEs, kcal/mol) at 298 K (BDE142/07): Theoretical errors [a] for various bond types

	Method	C–H (16)	X–H (12[b])	C–C (44)	C–O (19)	C–N (12)	C–F (5)	C–Cl (10)	C–S (9)	X–Y (15[c])
1st Rung	SPL	8.05	8.79	15.03	19.79	18.03	30.32	21.78	17.57	24.74
	SVWN	8.68	9.71	16.18	21.00	19.25	30.91	21.09	18.97	26.18
	SVWN5	7.79	8.39	14.07	18.78	17.13	28.54	22.09	16.34	23.87
2nd Rung	BLYP	4.18	4.95	10.97	10.17	10.55	2.57	4.94	8.08	6.28
	PBE	4.61	4.88	4.92	5.44	3.68	4.45	3.02	2.76	6.58
	BPW91	5.40	5.65	8.57	8.18	7.95	2.08	2.86	5.10	5.72
	BP86	2.75	2.83	6.77	6.44	5.76	3.07	2.61	3.66	5.91
3rd Rung	M06-L	4.21	4.40	5.03	5.53	6.62	2.00	2.89	3.54	3.42
	TPSS	3.27	3.91	9.72	8.21	9.62	2.38	2.85	5.61	4.91
	VSXC	3.45	3.78	6.50	6.98	7.66	2.03	2.13	4.59	5.11
4th Rung	BHHLYP	2.96	4.75	8.02	9.44	8.47	9.42	7.69	9.66	13.55
	B3PW91	4.44	3.64	7.16	6.06	6.28	1.89	3.06	4.97	4.15
	B3LYP	2.70	3.04	8.03	7.25	7.87	2.64	4.74	7.13	5.51
	B3P86	2.21	1.22	4.58	3.59	3.14	2.19	2.43	3.09	3.17
	PBE0	4.15	4.09	4.65	4.10	3.79	1.36	2.35	3.17	3.17
	BMK	1.44	1.99	2.71	2.44	1.15	1.64	1.72	1.30	2.87
	TPSSh	3.06	3.48	9.08	7.95	9.18	3.91	4.41	5.90	4.65
	M06-2X	1.49	2.31	1.41	3.17	2.16	2.02	2.87	1.06	2.95
	M06	2.09	1.38	2.74	1.67	1.79	3.13	2.76	2.41	2.76
5th Rung	XYG3	1.06	2.41	1.69	2.10	2.44	1.62	0.78	1.52	3.09
	MC3BB	2.18	1.66	4.11	1.32	1.98	4.05	4.06	1.04	1.45
	B2PLYP	2.19	2.87	4.15	3.61	4.25	1.60	2.18	3.73	3.55
	B2PLYP-D	1.98	2.71	2.65	2.46	2.86	1.57	1.70	2.72	3.04
	B2GP-PLYP	2.10	2.52	3.15	2.15	2.61	1.28	1.82	2.85	3.36
Ab initio	HF	23.75	26.85	31.23	32.95	30.91	42.85	33.14	32.49	44.75
	UMP2	5.44	4.28	11.28	7.47	9.18	13.95	12.50	3.73	5.24
	G2 [d]	1.53	1.42	2.51	3.50	2.60	3.35	2.93	1.41	1.65
	G3 [d]	0.50	1.35	1.23	2.12	1.32	0.99	1.30	0.85	2.11

[a] Mean absolute deviation, (MAD, kcal/mol). B3LYP/6–311 + G(d,p) optimized geometries are adopted for all calculations. ZPEs are scaled with 0.9877. B3LYP/6–311 + G(3df, 2p) is used for single point energy calculations [25–27]. Experimental data are from Refs. [8, 9].
[b] X–H: N–H, O–H, S–H, P–H, S–H.
[c] X–Y: O–O, N–N, N–O, C–Si, Si–Si, Cl–O, Cl–N.
[d] Data from Refs. [8, 9]. Calculated with the Gn theory [7–9]

satisfactory for C–O, with MADs of 3.50 and 2.12 kcal/mol, for G2 and G3, respectively.

Fifteen C–halogen bonds have been included in Table 3.5 for comparisons. G2 is particularly erroneous for these sets, with MADs of 3.35 and 2.93 kcal/mol for C–F and C–Cl, respectively. G3 improves over G2 by including the spin-orbital corrections for atoms [9]. Such corrections can be conveniently included in the

BDE calculations with DFAs. For these C–halogen bonds, MADs are reduced from B3LYP to B3PW91 to PBE0. These traditional hybrid functional perform better than the M06 family. The 3rd rung functionals are similarly good as the 4th rung functionals. With the exception of MC3BB, other DHDFs are good performers. For example, XYG3 gives MADs of 1.62 and 0.78 kcal/mol for C–F and C–Cl, respectively.

3.3.3 Basis Set Dependence of B3LYP, MP2 and XYG3

Figure 3.6 shows the basis set dependence for BDEs in the BDE92/07 set [10, 41]. For B3LYP, MAD from B1 = 6-311 + G(d, p) is 5.86 kcal/mol and that from B10 = 6-311 ++G(3df, 3pd) is 5.18 kcal/mol. Hence the basis set dependence is not evident. Basis set dependence for BDEs associated with MP2 is also significantly attenuated as compared to that for the corresponding HOFs, while it is still sizable as compared to that of BDEs associated with B3LYP. MAD from UMP2/B1 is 6.18 kcal/mol, which increases, rather than decreases, to 8.41 for UMP2/B10. This reflects that higher order correlation effects have to be introduced. XYG3 is particularly satisfying for BDE predictions. Even with XYG3/B1, MAD for the BDE92/07 set is 3.29 kcal/mol, smaller than the best values of UMP2/B1 (6.18) and B3LYP/B10 (5.17). Improving basis set improves steadily the XYG3 performance, such that XYG3/B10 gives MAD = 1.46 kcal/mol.

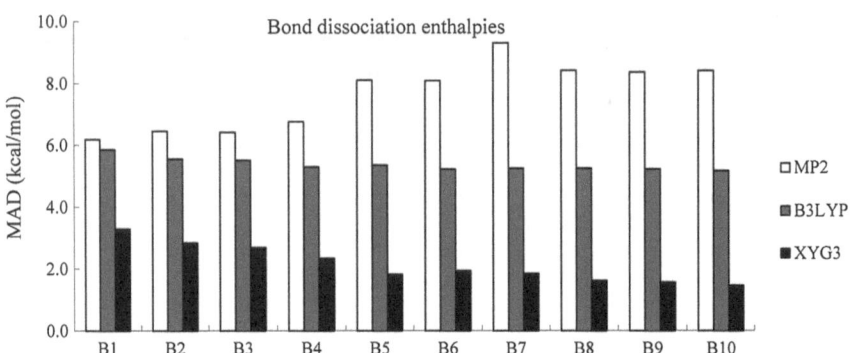

Fig. 3.6 Mean absolute deviations (MADs): basis set dependence for bond dissociation energies against the BDE92/07 set. B1: 6-311 + G(d, p), B2: 6-311 + G(2d, p), B3: 6-311 + G(2d, 2p), B4: 6-311 + G(3d, 2p), B5: 6-311 + G(2df, p), B6: 6-311 + G(2df, 2p), B7: 6-311 + G(3df), B8: 6-311 + G(3df, p), B9: 6-311 + G(3df, 2p), and B10: 6-311 ++G(3df, 3pd)

3.4 Reaction Barrier Heights Against the BH76 and PES(H-CH₄) Sets

3.4.1 Error Statistics Against the BH76 Set for Various DFT Methods

Zhao and Truhlar have compiled several benchmark sets for reaction barrier heights in 2004 [17, 23, 71–73]. These sets are grouped together and dubbed the name BH76/04. The functional performances summarized in Table 3.6 were based on the calculations with the 6-311 + G(3df, 2p) basis set [25–27].

Common DFAs are usually problematic for the stretched partially broken bonds due to the so-called self-interaction errors (SIE) [57]. This has led to a general tendency for common DFT methods to underestimate the reaction barrier heights. Table 3.6 shows that MAD is 15.18 kcal/mol for LDA(SPL), which is reduced to 9.04 and 8.22 kcal/mol for GGAs of PBE and BLYP, respectively, and is further

Table 3.6 Reaction barrier heights (RBHs, kcal/mol) for Truhlar's BH76/04 set: [a] Theoretical errors.[b]

	Method	UM10	NS16	HAT12	HT38	Total (76)
1st Rung	SPL	5.90	8.56	23.24	17.86	15.18
	SVWN	5.86	8.59	23.20	18.01	15.25
	SVWN5	6.01	8.46	23.44	17.78	15.16
2nd Rung	BLYP	3.57	7.33	14.69	7.77	8.22
	PBE	3.41	6.94	14.97	9.52	9.04
	BPW91	2.88	5.95	13.02	7.59	7.48
	BP86	3.94	6.87	15.55	9.38	9.11
3rd Rung	M06-L	1.86	3.33	5.86	4.36	4.05
	TPSS	4.08	7.84	14.67	7.93	8.47
	VSXC	2.45	5.01	7.49	5.12	5.12
4th Rung	BHHLYP	1.95	1.42	3.01	2.60	2.33
	B3LYP	2.02	3.38	8.51	4.43	4.54
	B3PW91	1.96	2.18	7.22	4.25	3.98
	B3P86	2.87	2.97	9.15	5.80	5.35
	X3LYP	2.06	3.41	8.51	4.57	4.62
	PBE0	2.23	1.99	6.66	4.44	3.99
	B97-1	0.94	1.38	1.47	1.27	1.28
	B97-D	3.14	6.02	9.72	7.32	6.87
	ωB97X	2.36	1.24	2.27	2.17	2.01
	ωB97X-D	1.89	0.80	2.05	2.31	1.90
	M06-2X	0.94	1.38	1.47	1.27	1.28
	M06	1.68	1.62	3.41	2.23	2.22
5th Rung	XYG3	0.98	1.42	1.38	0.75	1.02
	MC3BB	1.50	0.72	2.39	0.82	1.13
	B2PLYP	0.73	2.16	3.05	1.81	1.94

(continued)

Table 3.6 (continued)

	Method	UM10	NS16	HAT12	HT38	Total (76)
	B2PLYP-D	0.81	2.51	3.29	2.21	2.26
	B2GP-PLYP	0.98	0.90	1.86	0.73	0.97
Ab initio	HF	3.82	6.69	16.86	13.41	11.28
	UMP2	5.43	1.69	11.42	3.82	4.78
	QCISD(T)[c]	0.53	1.08	1.21	1.24	1.10

[a] BH76/04 set contains UM10, NS16, HAT12, and HT38 subsets, where HAT12 refers to the forward and reverse barrier heights for 6 heavy-atom transfer reactions, NS16 refers to the forward and reverse barrier heights for 8 nucleophilic substitution reactions, UM10 refers to the forward and reverse barrier heights for 5 association and unimolecular reactions, and HT38 refers to the forward and reverse barrier heights for 19 hydrogen transfer reactions.
[b] Mean absolute deviation, (MAD, kcal/mol). The calculations were performed using geometries from Truhlar database website [73]. W1 reference data are from Ref. [71].
[c] Data from Ref. [71]

reduced to 4.54 for the most widely used hybrid functional B3LYP. Hence, there is an improvement along the rungs up. The improvement with the third rung is, however, controversy. While M06-L gives MAD of 4.05 kcal/mol for this property, TPSS leads to MAD of 8.47 kcal/mol, which is comparable to those of GGAs.

Large portion of HF exchange proves to be valuable here. Doubled portion of HF exchange in M06-2X improves its performance over M06, with MAD of 1.28 kcal/mol for the former and 2.22 kcal/mol for the latter against the BH76 set.

Hybrid B97-1 functional [74] is also found to be very satisfactory in reaction barrier height calculations, giving MAD of only 1.28 kcal/mol for the BH76 set. Disturbingly, B97-D [75], which is a re-optimized version with the dispersion term is very poor for BH76 set, whose MAD is as high as 6.87 kcal/mol.

DHDFs generally represent an important advance. D2PLYP, MC3BB, XYG3, and B2GP-PLYP lead to MADs of 1.94, 1.13, 1.02 and 0.97 kcal/mol, respectively. This accuracy is comparable to that of the QCISD(T) ab initio method with the same basis set (MAD = 1.10 kcal/mol). We emphasize that barrier heights are not included in the XYG3 training set, but are included in the M06 family, MC3BB, and B2GP-PLYP training sets. Probably it is the presence of $\sim$ 80 % exact exchange in XYG3 that decreases the self-interaction errors (SIE) of local DFT functionals.

3.4.2 Potential Energy Curves for the $H + CH_4 \rightarrow H_2 + CH_3$ Reaction

Accurate potential energy surfaces (PES) are essential for using theory to predict chemical processes, but the accuracy depends critically on the level of the theory. Because of its important roles in CH_4/O_2 combustion chemistry, the $H + CH_4 \rightarrow H_2 + CH_3$ reaction has long been the subject of both experimental and theoretical interest [76].

Fig. 3.7 Comparison of the calculated potential energy curves for the $H + CH_4 \rightarrow H_2 + CH_3$ reaction with B3LYP, MP2 and XYG3 [24]. The CCSD(T)/6-311 ++G(3df, 2pd) data are used as the Ref. [76]. Reaction coordinate is defined as [R(CH)-R(HH)] (in Å)

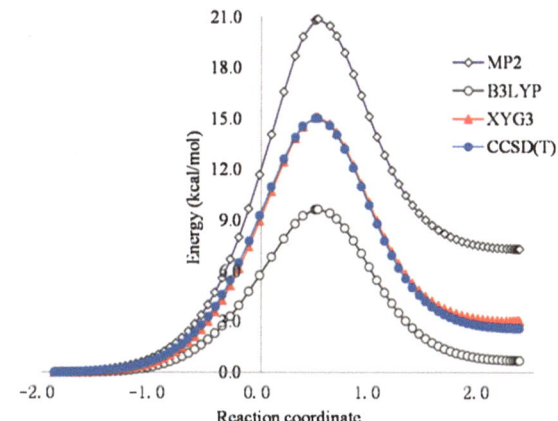

Figure 3.7 presents a point-to-point comparison among the results of various methods along the reaction coordinate. It is expected that the CCSD(T)/6-311 ++G(3df, 2pd) curve should be the most accurate, leading to a barrier of 15.03 kcal/mol. Remarkably XYG3 predicts the barrier of 15.08 kcal/mol, and is within 0.44 kcal/mol of the CCSD(T) results for the entire reaction path.

B3LYP and MP2 are both inadequate for the potential energy surface calculations. B3LYP underestimates the barrier height by 5.40 kcal/mol, whereas MP2 overestimates the barrier height by 5.81 kcal/mol. B3LYP and MP2 are also poor for reaction heats. While B3LYP underestimates it by 1.89 kcal/mol, MP2 overestimates the endothermicity of the reaction by 4.70 kcal/mol.

Figure 3.8 gives a comparison among other DHDFs and CCST(T) results. MC3BB is less satisfactory for this reaction. The maximum error is as high as 2.85 kcal/mol, occurring at the product area. As comparing the MC3BB curve with the MP2 curve, it is clear that MC3BB inherits some weakness of MP2. B2PLYP

Fig. 3.8 Comparison of the calculated potential energy curves for the $H + CH_4 \rightarrow H_2 + CH_3$ reaction with MC3BB, B2PLYP and B2GP-PLYP [10]. The CCSD(T)/6-311 ++G(3df, 2pd) data are used as the Ref. [76]. Reaction coordinate is defined as [R(CH)-R(HH)] (in Å)

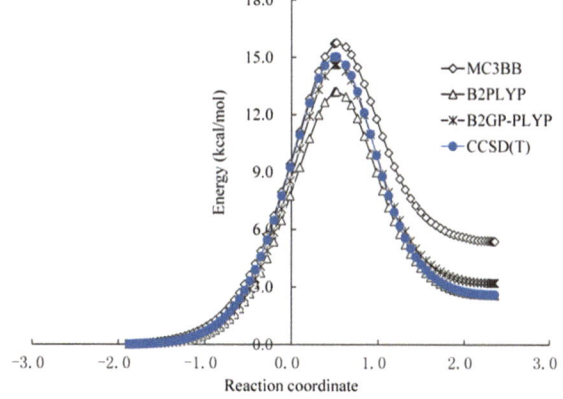

leads to very good reaction heat for this reaction (2.58 as compared to 2.59 kcal/mol of CCSD(T)). However, B2PLYP underestimates the barrier height by 1.83 kcal/mol. It is shown that the B2GP-PLYP functional is quite satisfactory for the description of this potential energy curve. The barrier height is 0.41 kcal/mol lower and the product level is 0.61 kcal/mol higher than the corresponding CCSD(T) values.

3.4.3 Basis Set Dependence for the Calculated Reaction Barrier Heights

Figure 3.9 displays the results of basis set dependence against BH76 as comparing the performance of B3LYP, MP2, and XYG3 with basis sets from B1 = 6-311 + G(d, p) to B10 = 6311 ++G(3df, 3pd) [25–27]. The basis set dependence is found to be quite mild for B3LYP. This is also true for MP2 and XYG3, showing that triple-zeta basis set plus a minimum set of polarization functions is generally good for barrier height predictions. B3LYP with various basis sets leads to MAD around 4.6 kcal/mol. MAD from MP2/B1 is 5.82 kcal/mol, which is reduced to 4.44 kcal/mol with B10. XYG3 is obviously superior to B3LYP and MP2, leading to MAD around 2.0 kcal/mol for B1, B3, B4, and B7 and around 1.0 kcal/mol for other basis sets.

We have also studied the basis set dependence for the calculated potential energy curves with XYG3. Figure 3.10 shows such a comparison. The B7 = 6-311 + G(3df) results is not included as it contains no polarization function on hydrogen atoms, which it is certainly unbalanced for this H-abstraction system. The close agreement among the results of various basis sets is encouraging, which once more demonstrates quite good basis set convergence for the XYG3 functional.

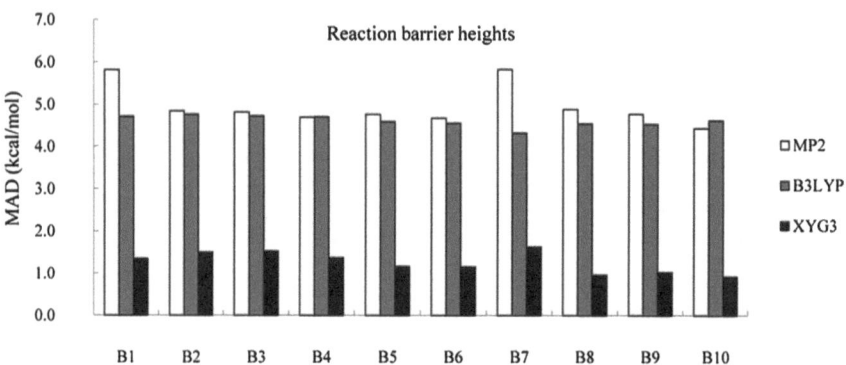

Fig. 3.9 Mean absolute deviations (MADs): basis set dependence for reaction barrier heights against the BH76/04 set

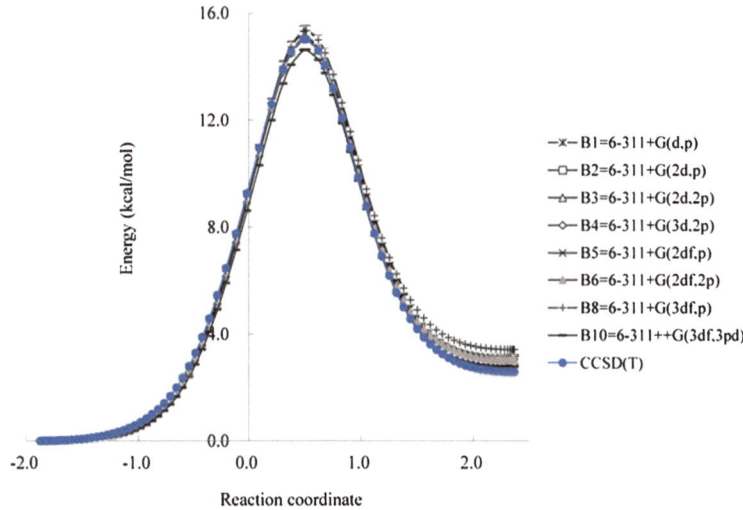

Fig. 3.10 Basis set dependence for the calculated potential energy curves of the $H + CH_4 \rightarrow H_2 + CH_3$ reaction [41]

3.5 Nonbonded Interactions Against the NCIE31 and PES(CH₄-C₆H₆) Sets

3.5.1 Error Statistics Against the NCIE31 Set for Various DFT Methods

Noncovalent interactions (NCIs) determine the structures of biomacromolecules like DNA, RNA, and proteins [77–83]. To elucidate their geometric and electronic structures, reliable calculations are mandatory [83–87]. Zhao and Truhlar have constructed a set, dubbed NCIE31/05, which has now been frequently used as a training and/or testing set of NCIs for the development of DFAs [17, 23, 71–73]. This set covers various kinds of NCIs, consisting of 6 hydrogen bond (HB) complexes, 7 charge-transfer (CT) complexes, 6 dipole interaction (DI) complexes, 7 weak interaction (WI) complexes, and 5 $\pi-\pi$ stacking (PPS) complexes.

The errors against NCIE31/05 for various DFT methods are summarized in Table 3.7. The basis set used is generally 6-311 + G(3df, 2p). Basis set superposition error corrections are not included, which may facilitate the NCI calculations of intramolecular interactions. Geometries and reference energies are taken from the Truhlar database website without modifications [73].

LDA(SPL [11, 56]) leads to a relatively good performance for WI7 (MAD = 0.34 kcal/mol) and PPS5 (MAD = 0.68 kcal/mol), but is particularly poor for HB6, CI7, and DI6, with MADs of 4.61, 6.83, and 3.13 kcal/mol, respectively. The final MAD for the whole NCIE31/05 set is as high as 3.23 kcal/mol.

Table 3.7 Noncovalent interaction energies (kcal/mol) for Truhlar's NCIE31/05[a] set:theoretical errors[b]

	Method	HB6	CT7	DI6	WI7	PPS5	Total (31)
1st Rung	SPL	4.61	6.83	3.13	0.34	0.68	3.23
	SVWN	4.85	7.03	3.27	0.36	0.76	3.36
	SVWN5	4.60	6.82	3.11	0.34	0.68	3.22
2nd Rung	BLYP	1.20	1.71	0.85	0.41	3.58	1.45
	PBE	0.41	3.01	0.56	0.16	1.84	1.20
	BPW91	1.67	1.44	0.98	0.67	3.79	1.60
	BP86	0.75	2.10	0.64	0.61	3.18	1.39
3rd Rung	M06-L	0.26	1.96	0.65	0.20	0.20	0.70
	TPSS	0.43	2.24	0.50	0.20	2.48	1.13
	VSXC	0.35	2.94	1.31	1.01	7.26	2.38
4th Rung	B3LYP	0.63	0.76	0.61	0.27	2.93	0.95
	B3PW91	1.06	0.67	0.77	0.49	3.16	1.13
	B3P86	0.35	1.13	0.48	0.43	2.51	0.92
	X3LYP	0.41	1.03	0.38	0.17	2.47	0.82
	PBE0	0.40	1.13	0.36	0.13	1.79	0.72
	B97-1	0.37	1.26	0.36	0.11	1.56	0.70
	B97-D	0.37	1.91	0.42	0.18	0.20	0.66
	ωB97X	0.87	0.74	0.80	0.05	0.44	0.57
	ωB97X-D	0.37	0.33	0.38	0.06	0.58	0.33
	M06-2X	0.39	0.43	0.29	0.20	0.22	0.31
	M06	0.22	1.12	0.45	0.23	0.18	0.46
5th Rung	XYG3	0.38	0.64	0.19	0.12	0.26	0.32
	MC3BB	0.64	0.27	0.59	0.24	1.06	0.52
	B2PLYP	0.35	0.75	0.30	0.12	1.30	0.53
	B2PLYP-D	0.62	1.17	0.48	0.13	0.28	0.55
	B2GP-PLYP	0.35	0.70	0.23	0.09	0.77	0.41
Ab initio	HF	2.25	3.61	2.17	0.29	3.32	2.27
	UMP2	0.99	0.47	0.29	0.08	1.69	0.64
	QCISD(T) [d]	0.90	0.62	0.47	0.07	0.95	0.57

[a] NCIE31/05 set consists of 6 hydrogen bond (HB) complexes, 7 charge-transfer (CT) complexes, 6 dipole interaction (DI) complexes, 7 weak interaction (WI) complexes, and 5 $\pi - \pi$ stacking (PPS) complexes. [b] Mean absolute deviation, (MAD, kcal/mol). Our calculations used the 6-311 + G(3df, 2p) basis sets with geometries from Truhlar database website [73]. Counterpoise corrections [90] for possible basis set superposition errors were not included. W1 reference data are from Ref. [72]

   GGAs such as BLYP [13, 21] and PBE [15] lead to MAD of 1.45 and 1.20 kcal/mol for the same set, showing a good improvement in general. PBE is quite satisfactory for HB6, DI6, and WI7 (MADs = 0.41, 0.56, and 0.16 kcal/mol, respectively), but is pretty bad for CT7 and PPS5 (MADs = 3.01 and 1.84 kcal/mol, respectively). TPSS [16] (MAD = 1.13 kcal/mol) improves over PBE slightly, while M06-L [17] (MAD = 0.70 kcal/mol) displays a clear improvement.
   The M06 family of functionals is indeed outstanding, giving MADs of 0.31, and 0.46 kcal/mol for M06-2X, and M06, respectively [23]. In particular, M06-2X and

M06 give MADs of 0.22 and 0.18 kcal/mol, respectively, for the PPS5 subset, where the traditional hybrid functionals such as B3LYP [13, 19–21] (MAD = 2.93 kcal/mol) and PBE0 [15, 58, 64] (MAD = 1.79 kcal/mol) find this dispersion dominant subset extremely difficult.

It has been argued that it is due to an artifactual exchange effect that the traditional functionals may give dispersion-like binding near minimum-energy separations [88]. Nevertheless, dispersion is a dynamical correlation effect, arising from instantaneous fluctuation of the electron density [88, 89]. Even though some nonlocal information is introduced by including the kinetic-energy density, meta-GGAs still do not incorporate the physics of dispersion correlations [88].

Conventional DFAs miss the R^{-6} decay behavior in the long-range correlation [86]. The delocalized PT2 term in DHDFs, on the other hand, captures the correct long-range behavior by construction. Hence, XYG3 leads to MAD of 0.32 kcal/mol for the NCIE31/05 set, showing significant improvement over B3LYP on all its five subsets.

Note that these nonbonded interactions were not included in the XYG3 training set, but were included in the M06 training set. The performances of other DHDFs are comparable to that of XYG3 for other subsets with the exception of the PPS subset (see Table 3.7). It has been argued that it is the type of the orbitals used to evaluate the PT2 term that makes such difference among different DHDFs [10, 24]. Generally, only ∼ 30 % PT2 correlation is used to replace part of the DFT correlation. This PT2 portion is too small to embrace the full pieces of long-range dispersion [91].

There are other empirical and nonempirical approaches to modeling dispersion interactions within DFT methods [92–98]. Indeed, as shown in Table 3.7, B97-D and B2PLYP-D significantly improve over B97-1 and B2PLYP for the description of the PPS subset. These –D methods have imposed the correct long-range R^{-6} interatomic dependence by adding the functionals a posterior force field (FF) like dispersion correction. While such DFT-D schemes can be implemented efficiently without additional computational cost, it includes many empirical parameters, and suffers from some inherent limitations [94, 95, 99]. Especially, the many-body correlation effects and anisotropy effects in the long-range dispersive interactions, as well as the orbital-dependence in the medium range [30, 69, 95, 96, 99, 100], are more subtle, which are difficult to be approximated in the pair-wise additive FF models.

3.5.2   Intermolecular Potential Curves for the CH$_4$-C$_6$H$_6$ Complex

As steric constraints might prevent the ligand from adopting its optimum geometry, proper description of the whole potential energy curve is very important for describing the binding of ligands to biological systems. Figure 3.11 compares the intermolecular potentials of the CH$_4$-C$_6$H$_6$ complex calculated by B3LYP, MP2, and XYG3 [24]. The CCSD(T) results at the complete basis set (CBS) limit are used as Ref. [101].

Fig. 3.11 The
intermolecular potentials for
the CH_4-C_6H_6 complexes
from B3LYP, MP2, and
XYG3. R is defined as carbon
of CH_4 to the ring center of
C_6H_6 (in Å). Data in solid
blue are CCSD(T) at the
complete basis set limit from
Ref. [101]. A pruned (75,302)
grid is used for B3LYP and
XYG3

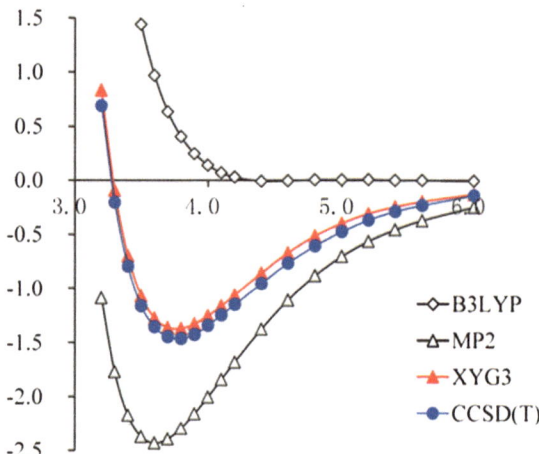

As shown in Fig. 3.11, B3LYP displays a repulsive potential energy curve, whereas MP2 leads to a potential energy curve which is too attractive. The XYG3 results compare very well with those of CCSD(T), where the deviations are generally smaller than 0.1 kcal/mol. It should be emphasized that basis set superposition error (BSSE [90]) corrections were not included in all these calculations. This omission, although may facilitate real applications, can deteriorate the statistics for some functionals, as well as MP2, if their potential energy curve is overbonded, or if they were developed when BSSE corrections were taken into account. The BSSE effects of XYG3 have been checked by Sherrill and co-workers [102]. The counterpoise-corrected XYG3 curve was found to be under-bound, and it was suggested to use the uncorrected results for XYG3.

Figure 3.12 plots the intermolecular potentials for other DHDFs. It is an obvious advancement that all these fifth rung functionals correctly lead to appreciable attractive wells, showing the importance to include the delocalized PT2 correlations. Quantitatively, however, all these DHDFs underestimate the binding energies. B2PLYP gives only half of the binding energy, MC3BB and B2GP-PLYP recover two-third of the binding energy as compared to the CCSD(T) reference value. Errors become quite significant in the wall area. For example, at R = 3.2 Å, MC3BB, B2GP-PLYP, and B2PLYP are too repulsive by 0.78, 0.95, and 1.64 kcal/mol, respectively, as compared to the CCSD(T)/CBS value.

It was found that meta-GGA functionals demand a very large integration grid to avoid spurious oscillations on potential energy curves for dispersion-bound complexes [88]. Wiggles are still detectible for MC3BB which uses B95 [103] meta-GGA correlation functional, even though a fine unpruned (250,590) grid has already been adopted in the calculations. Similar problems were encountered for the M06 family of functionals.

All other DFA calculations listed in Figs. 3.11 and 3.12 used a cheaper pruned (75,302) grid. In fact, meta-GGA calculations with finer grid can be more expensive than the PT2 calculations. For the CH_4-C_6H_6 complex with the

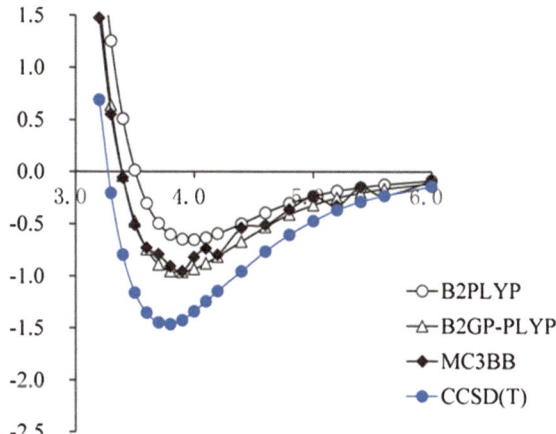

Fig. 3.12 The intermolecular potentials for the CH_4-C_6H_6 complexes from MC3BB, B2PLYP and B2GP-PLYP. R is defined as carbon of CH_4 to the ring center of C_6H_6 (in Å). Data in solid blue are CCSD(T) at the complete basis set limit from Ref. [101]. An unpruned (250,590) grid is used in calculations with MC3BB, while a pruned (75,302) grid is used for other DHDFs

6-311 + G(3df,2p) basis set, it takes about 7 min to run a M06-L calculation using the pruned (75,302) grid for one point on the potential curve with a dual processor Intel Xeon 5345–3.33 GHz Clovertown computer. It increases to ~11 and ~40 min with (99,590) and (250,590) grids, respectively, while XYG3 takes about 25 min job cpu time [10].

Basis set dependence of XYG3 on the calculated intermolecular potentials of the CH_4-C_6H_6 complex has also been investigated [41]. The results are reproduced in Fig. 3.13. The basis set dependence is not very significant for this system. All basis sets from B1 = 6-311 + G(d, p) to B10 = 6-311 ++G(3df, 3pd) lead to good results, correctly predicting the equilibrium geometry. The largest deviation (0.44 kcal/mol) occurs at the wall area with basis set of B2 = 6-311 + G(2d, p). XYG3/B1 seems to be of practical value for NCIs, partly benefiting from BSSE.

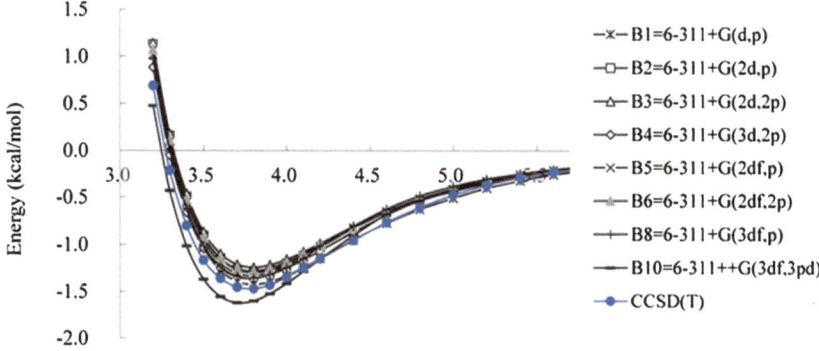

Fig. 3.13 Basis set dependence for the calculated intermolecular potentials of the CH_4-C_6H_6 complexes [41]. We omit the B7 results as it contains no polarization function on hydrogen atoms, and hence is unbalanced. Results for B9 are also not shown here, as B9 is the default 6-311 + G(3df, 2p) basis set used in this work

References

1. Lide DR (2001) CRC handbook of chemistry and physics, 84th edn. CRC Press, Boca Raton
2. Callonion JH, Hirota E, Kuchitsu K, Lafferty WJ, Maki AG (1976) Numerical data and function relationships in science and technology. Springer, West Berlin
3. Masterton ML, Slowinski EJ, Stanitski CL (1983) Chemical principles. CBS College Publishing, Philadelphia
4. Neutral Thermochemical Data (2005) NIST Chemistry WebBook, http://webook.nist.gov/chemistry. Accessed 15 Aprl 2013
5. Chase MW, Davies CA, Downey JR et al (1985) Janaf thermochemical tables—3rd edition.Parts 1 (Al-Co). J Phys Chem Ref Data 14:1–926. doi:10.1063/1.555747
6. Ruscic B, Boggs JE, Burcat A et al (2005) IUPAC critical evaluation of thermochemical properties of selected radicals. Part I. J Phys Chem Ref Data 34:573–656. doi:10.1063/1.1724828
7. Pople JA, Head-Gordon M, Fox DJ et al (1989) Gaussian-1 theory - A general procedure for prediction of molecular-energies. J Chem Phys 90:5622–5629. doi:10.1063/1.456415
8. Curtiss LA, Raghavachari K, Trucks GW, Pople JA (1991) Gaussian-2 theory for molecular-energies of 1st-row and 2nd-row compounds. J Chem Phys 94:7221–7230. doi:10.1063/1.460205
9. Curtiss LA, Raghavachari K, Redfern PC et al (1998) Gaussian-3 (G3) theory for molecules containing first and second-row atoms. J Chem Phys 109:7764–7776. doi:10.1063/1.477422
10. Zhang IY, Xu X (2011) Doubly hybrid density functional for accurate description of thermochemistry, thermochemical kinetics and nonbonded interactions. Int Rev Phys Chem 30:115–160. doi:10.1080/0144235X.2010.542618
11. Slater JC (1960) Quantum theory of atomic structure, vol 2. McGraw-Hill, New York
12. Vosko SH, Wilk L, Nusair M (1980) Accurate spin-dependent electron liquid correlation endergies for local spin-density calculations–a critical analysis. Can J Phys 58:1200–1211. doi:10.1139/p80-159
13. Becke AD (1988) Density-functional exchange-energy approximation with correct asymptotic behavior. Phys Rev A 38:3098–3100. doi:10.1103/PhysRevA.38.3098
14. Perdew JP, Chevary JA, Vosko SH et al (1992) Atoms, molecules, solids, and surfaces: Applications of the generalized gradient approximation for exchange and correlation. Phys Rev B 46:6671–6687. doi:10.1103/PhysRevB.46.6671
15. Perdew JP, Burke K, Ernzerhof M (1996) Generalized gradient approximation made simple. Phys Rev Lett 77:3865–3868. doi:10.1103/PhysRevLett.77.3865
16. Tao JM, Perdew JP, Staroverov VN, Scuseria GE (2003) Climbing the density functional ladder: Nonempirical meta-generalized gradient approximation designed for molecules and solids. Phys Rev Lett 91:146401. doi:10.1103/PhysRevLett.91.146401
17. Zhao Y, Truhlar DG (2006) A new local density functional for main-group thermochemistry, transition metal bonding, thermochemical kinetics, and noncovalent interactions. J Chem Phys 125:194101. doi:10.1063/1.2370993
18. Van Voorhis T, Scuseria GE (1998) A novel form for the exchange-correlation energy functional. J Chem Phys 109:400–410. doi:10.1063/1.476577
19. Becke AD (1993) Density-functional thermochemistry. 3. The role of exact exchange. J Chem Phys 98:5648–5652. doi:10.1063/1.464913
20. Stephens PJ, Devlin FJ, Chabalowski CF, Frisch MJ (1994) Ab-initio calculation of vibrational absorption and circular-dichroism spectra using density-functional force-fields. J Phys Chem 98:11623–11627. doi:10.1021/j100096a001
21. Lee CT, Yang WT, Parr RG (1988) Development of the Colle–Salvetti correlation-energy formula into a functional of the electron-density. Phys Rev B 37:785–789. doi:10.1103/PhysRevB.37.785

22. Chai J-D, Head-Gordon M (2008) Long-range corrected hybrid density functionals with damped atom–atom dispersion corrections. Phys Chem Chem Phys 10:6615–6620. doi:10.1039/B810189B

23. Zhao Y, Truhlar DG (2008) The M06 suite of density functionals for main group thermochemistry, thermochemical kinetics, noncovalent interactions, excited states, and transition elements: two new functionals and systematic testing of four M06-class functionals and 12 other functionals. Theor Chem Acc 120:215–241. doi:10.1007/s00214-007-0310-x

24. Zhang Y, Xu X, Goddard WA (2009) Doubly hybrid density functional for accurate descriptions of nonbond interactions, thermochemistry, and thermochemical kinetics. Proc Natl Acad Sci USA 106:4963–4968. doi:10.1073/pnas.0901093106

25. Frisch MJ et al. (2003) Gaussian 03, revision A. 1. Gaussian, Inc, Pittsburgh

26. Krishnan R, Binkley JS, Seeger R, Pople JA (1980) Self-consistent molecular orbital methods. XX. A basis set for correlated wave functions. J Chem Phys 72:650–654. doi:10.1063/1.438955

27. Frisch MJ, Pople JA, Binkley JS (1984) Self–consistent molecular orbital methods 25. Supplementary functions for Gaussian basis sets. J Chem Phys 80:3265–3269. doi:10.1063/1.447079

28. Zhao Y, Lynch BJ, Truhlar DG (2004) Doubly hybrid meta DFT: New multi-coefficient correlation and density functional methods for thermochemistry and thermochemical kinetics. J Phys Chem A 108:4786–4791. doi:10.1021/jp049253v

29. Grimme S (2006) Semiempirical hybrid density functional with perturbative second-order correlation. J Chem Phys 124:034108–034116. doi:10.1063/1.2148954

30. Schwabe T, Grimme S (2007) Double-hybrid density functionals with long-range dispersion corrections: higher accuracy and extended applicability. Phys Chem Chem Phys 9:3397–3406. doi:10.1039/b704725h

31. Weigend F, Ahlrichs R (2005) Balanced basis sets of split valence, triple zeta valence and quadruple zeta valence quality for H to Rn: Design and assessment of accuracy. Phys Chem Chem Phys 7:3297–3305. doi:10.1039/B508541A

32. Sharkas K, Savin A, Jensen HJA, Toulouse J (2012) A multiconfigurational hybrid density-functional theory. J Chem Phys 137:044104. doi:10.1063/1.4733672

33. Langreth DC, Perdew JP (1977) Exchange-correlation energy of a metallic surface: Wave-vector analysis. Phys Rev B 15:2884–2901. doi:10.1103/PhysRevB.15.2884

34. Becke AD (1993) A new mixing of Hartree–Fock and local density-functional theories. J Chem Phys 98:1372–1377. doi:10.1063/1.464304

35. Gunnarsson O, Lundqvist BI (1976) Exchange and correlation in atoms, molecules, and solids by the spin-density-functional formalism. Phys Rev B 13:4274–4298. doi:10.1103/PhysRevB.13.4274

36. Perdew JP, Emzerhof M, Burke K (1996) Rationale for mixing exact exchange with density functional approximations. J Chem Phys 105:9982–9985. doi:10.1063/1.472933

37. Mori-Sánchez P, Cohen AJ, Yang WT (2006) Self-interaction-free exchange-correlation functional for thermochemistry and kinetics. J Chem Phys 124:091102. doi:10.1063/1.2179072

38. Görling A, Levy M (1993) Correlation-energy functional and its high-density limit obtained from a coupling-constant perturbation expansion. Phys Rev B 47:13105–13113. doi:10.1103/PhysRevB.47.13105

39. Cremer D (2001) Density functional theory: coverage of dynamic and non-dynamic electron correlation effects. Mol Phys 99:1899–1940. doi:10.1080/00268970110083564

40. Wu JM, Xu X (2007) The X1 method for accurate and efficient prediction of heats of formation. J Chem Phys 127:214105–214113. doi:10.1063/1.2800018

41. Zhang I, Luo Y, Xu X (2010) Basis set dependence of the doubly hybrid XYG3 functional. J Chem Phys 133:104105. doi:10.1063/1.3488649

42. Boese A, Martin J, Handy NC (2003) The role of the basis set: Assessing density functional theory. J Chem Phys 119:3005–3014. doi:10.1063/1.1589004

43. Curtiss LA, Raghavachari K, Redfern PC, Pople JA (2000) Assessment of Gaussian-3 and density functional theories for a larger experimental test set. J Chem Phys 112:7374–7383. doi:10.1063/1.481336
44. Job G, Herrmann F (2006) Chemical potential—A quantity in search of recognition. Eur J Phys 27:353. doi:10.1088/0143-0807/27/2/018
45. Berkowitz M, Parr RG (1988) Molecular hardness and softness, local hardness and softness, hardness and softness kernels, and relations among these quantities. J Chem Phys 88:2554–2557. doi:10.1063/1.454034
46. Ingold CK (1934) Principles of an electronic theory of organic reactions. Chem Rev 15:225–274. doi:10.1021/cr60051a003
47. Mulliken RS (1934) A new electroaffinity scale; together with data on valence states and on valence ionization potentials and electron affinities. J Chem Phys 2:782–793. doi:10.1063/1.1749394
48. Yokojima S, Yoshiki N, Yanoi W, Okada A (2009) Solvent effects on ionization potentials of guanine runs and chemically modified guanine in duplex DNA: Effect of electrostatic interaction and its reduction due to solvent. J Phys Chem B 113:16384–16392. doi:10.1021/jp9054582
49. Steenken S, Telo JP, Novais HM, Candeias LP (1992) One-electron-reduction potentials of pyrimidine bases, nucleosides, and nucleotides in aqueous solution. Consequences for DNA redox chemistry. J Am Chem Soc 114:4701–4709. doi:10.1021/ja00038a037
50. Khistyaev K, Bravaya KB, Kamarchik E et al (2011) The effect of microhydration on ionization energies of thymine. Faraday Discuss 150:313–330. doi:10.1039/C0FD00002G
51. Vijayaraj R, Subramanian V, Chattaraj PK (2009) Comparison of global reactivity descriptors calculated using various density functionals: A QSAR perspective. J Chem Theory Comput 5:2744–2753. doi:10.1021/ct900347f
52. Fayet G, Joubert L, Rotureau P, Adamo C (2009) On the use of descriptors arising from the conceptual density functional theory for the prediction of chemicals explosibility. Chem Phys Lett 467:407–411. doi:10.1016/j.cplett.2008.11.033
53. Geerlings P, De Proft F, Langenaeker W (2003) Conceptual density functional theory. Chem Rev 103:1793–1873. doi:10.1021/cr990029p
54. Thanikaivelan P, Subramanian V, Raghava Rao J, Unni Nair B (2000) Application of quantum chemical descriptor in quantitative structure activity and structure property relationship. Chem Phys Lett 323:59–70. doi:10.1016/S0009-2614(00)00488-7
55. Su NQ, Zhang IY, Wu JM, Xu X (2011) Calculations of ionization energies and electron affinities for atoms and molecules: A comparative study with different methods. Front Chem China 6:269–279. doi:10.1007/s11458-011-0256-3
56. Perdew JP, Zunger A (1981) Self-interaction correction to density-functional approximations for many-electron systems. Phys Rev B 23:5048–5079. doi:10.1103/PhysRevB.23.5048
57. Cohen AJ, Mori-Sánchez P, Yang WT (2011) Challenges for density functional theory. Chem Rev 112:289–320. doi:10.1021/cr200107z
58. Ernzerhof M, Scuseria GE (1999) Assessment of the Perdew-Burke-Ernzerhof exchange-correlation functional. J Chem Phys 110:5029–5036. doi:10.1063/1.478401
59. Galbraith JM, Schaefer HF (1996) Concerning the applicability of density functional methods to atomic and molecular negative ions. J Chem Phys 105:862–864. doi:10.1063/1.471933
60. Rösch N, Trickey SB (1997) Concerning the applicability of density functional methods to atomic and molecular negative ions–Comment. J Chem Phys 106:8940–8941. doi:10.1063/1.473946
61. Wu JM, Xu X (2008) Improving the B3LYP bond energies by using the X1 method. J Chem Phys 129:164103–164111. doi:10.1063/1.2998231
62. Zhang IY, Wu J, Luo Y, Xu X (2010) Trends in R − X Bond dissociation energies (R· = Me, Et, i-Pr, t-Bu, X· = H, Me, Cl, OH). J Chem Theory Comput 6:1462–1469. doi:10.1021/ct100010d
63. Zhang IY, Wu J, Luo Y, Xu X (2011) Accurate bond dissociation enthalpies by using doubly hybrid XYG3 functional. J Comput Chem 32:1824–1838. doi:10.1002/jcc.21764

64. Adamo C, Barone V (1999) Toward reliable density functional methods without adjustable parameters: The PBE0 model. J Chem Phys 110:6158–6170. doi:10.1063/1.478522
65. Karton A, Tarnopolsky A, Lamère JF et al (2008) Highly accurate first-principles benchmark data sets for the parametrization and validation of density functional and other approximate methods. Derivation of a robust, generally applicable, double-hybrid functional for thermochemistry and thermochemical kinetics. J Phys Chem A 112:12868–12886. doi:10.1021/jp801805p
66. Coote ML (2004) Reliable theoretical procedures for the calculation of electronic-structure information in hydrogen abstraction reactions. J Phys Chem A 108:3865–3872. doi:10.1021/jp049863v
67. Izgorodina E, Coote M, Radom L (2005) Trends in R-X bond dissociation energies (R = Me, Et, i-Pr, t-Bu; X = H, CH3, OCH3, OH, F): A surprising shortcoming of density functional theory. J Phys Chem A 109:7558–7566. doi:10.1021/jp052021r
68. Check C, Gilbert T (2005) Progressive systematic underestimation of reaction energies by the B3LYP model as the number of C–C bonds increases: Why organic chemists should use multiple DFT models for calculations involving polycarbon hydrocarbons. J Org Chem 70:9828–9834. doi:10.1021/jo051545k
69. Grimme S (2006) Seemingly simple stereoelectronic effects in alkane isomers and the implications for Kohn-Sham density functional theory. Angew Chem Int Ed 45:4460–4464. doi:10.1002/anie.200600448
70. Wodrich MD, Corminboeuf C, Schleyer PV (2006) Systematic errors in computed alkane energies using B3LYP and other popular DFT functionals. Org Lett 8:3631–3634. doi:10.1021/ol061016i
71. Zhao Y, González-García N, Truhlar DG (2005) Benchmark database of barrier heights for heavy atom transfer, nucleophilic substitution, association, and unimolecular reactions and its use to test theoretical methods. J Phys Chem A 109:2012–2018. doi:10.1021/jp045141s
72. Zhao Y, Truhlar DG (2005) Design of density functionals that are broadly accurate for thermochemistry, thermochemical kinetics, and nonbonded interactions. J Phys Chem A 109:5656–5667. doi:10.1021/jp050536c
73. Minnesota Database Collection (2006) Lynch BJ, Zhao Y, Truhlar DG. http://t1.chem.umn.edu/misc/database_group/database_therm_bh. Accessed 15 Aprl 2013
74. Hamprecht FA, Cohen AJ, Tozer DJ, Handy NC (1998) Development and assessment of new exchange-correlation functionals. J Chem Phys 109:6264–6271. doi:10.1063/1.477267
75. Grimme S (2006) Semiempirical GGA-type density functional constructed with a long-range dispersion correction. J Comput Chem 27:1787–1799. doi:10.1002/jcc.20495
76. Zhang LL, Lu YP, Lee SY, Zhang DH (2007) A transition state wave packet study of the H + CH4 reaction. J Chem Phys 127:234313. doi:10.1063/1.2812553
77. Saenger W (1984) Principles of nucleic acid structure. Springer, New York
78. Burley SK, Petsko GA (1985) Aromatic-aromatic interaction—A mechanism of protein-structure stabilization. Science 229:23–28. doi:10.1126/science.3892686
79. Lehn J-M (1990) Perspectives in supramolecular chemistry—From molecular recognition towards molecular information-processing and self-organization. Angew Chem Int Ed 29:1304–1319. doi:10.1002/anie.199013041
80. Guallar V, Borrelli KW (2005) A binding mechanism in protein-nucleotide interactions: Implication for U1A RNA binding. Proc Natl Acad Sci USA 102:3954–3959. doi:10.1073/pnas.0500888102
81. Vondrášek J, Bendová L, Klusák V, Hobza P (2005) Unexpectedly strong energy stabilization inside the hydrophobic core of small protein rubredoxin mediated by aromatic residues: correlated ab initio quantum chemical calculations. J Am Chem Soc 127:2615–2619. doi:10.1021/ja044607h
82. Dąbkowska I, Gonzalez HV, Jurečka P, Hobza P (2005) Stabilization energies of the hydrogen-bonded and stacked structures of nucleic acid base pairs in the crystal geometries of CG, AT, and AC DNA steps and in the NMR geometry of the 5'-d(GCGAAGC)-3'

hairpin: Complete basis set calculations at the MP2 and CCSD(T) levels RID A-6885-2008. J Phys Chem A 109:1131–1136. doi:10.1021/jp046738a

83. Müller-Dethlefs K, Hobza P (2000) Noncovalent interactions: A challenge for experiment and theory. Chem Rev 100:143–167. doi:10.1021/cr9900331

84. Hobza P, Šponer J (1999) Structure, energetics, and dynamics of the nucleic acid base pairs: Nonempirical ab initio calculations. Chem Rev 99:3247–3276. doi:10.1021/cr9800255

85. Sinnokrot MO, Valeev EF, Sherrill CD (2002) Estimates of the ab initio limit for pi–pi interactions: The benzene dimer. J Am Chem Soc 124:10887–10893. doi:10.1021/ja025896h

86. Riley KE, Pitoňák M, Jurečka P, Hobza P (2010) Stabilization and structure calculations for noncovalent interactions in extended molecular systems based on wave function and density functional theories. Chem Rev 110:5023–5063. doi:10.1021/cr1000173

87. Morgado CA, Jurečka P, Svozil D et al (2010) Reference MP2/CBS and CCSD(T) quantum-chemical calculations on stacked adenine dimers. Comparison with DFT-D, MP2.5, SCS(MI)-MP2, M06–2X, CBS(SCS-D) and force field descriptions. Phys Chem Chem Phys 12:3522–3534. doi:10.1039/b924461a

88. Johnson ER, Becke AD, Sherrill CD, DiLabio GA (2009) Oscillations in meta-generalized-gradient approximation potential energy surfaces for dispersion-bound complexes. J Chem Phys 131:034111–034117. doi:10.1063/1.3177061

89. Dobson JF, McLennan K, Rubio A et al (2001) Prediction of dispersion forces: Is there a problem. Aust J Chem 54:513–527. doi:10.1071/CH01052

90. Boys SF, Bernardi F (2002) The calculation of small molecular interactions by the differences of separate total energies. Some procedures with reduced errors (Reprinted from Molecular Physics, vol 19, pg 553–566, 1970). Mol Phys 100:65–73. doi:10.1080/00268970110088901

91. Goerigk L, Grimme S (2011) Efficient and accurate double-hybrid-meta-GGA density functionals—evaluation with the extended GMTKN30 database for general main group thermochemistry, kinetics, and noncovalent interactions. J Chem Theory Comput 7:291–309. doi:10.1021/ct100466k

92. Wu Q, Yang WT (2002) Empirical correction to density functional theory for van der Waals interactions. J Chem Phys 116:515–524. doi:10.1063/1.1424928

93. Grimme S (2004) Accurate description of van der Waals complexes by density functional theory including empirical corrections. J Comput Chem 25:1463–1473. doi:10.1002/jcc.20078

94. Grimme S, Antony J, Ehrlich S, Krieg H (2010) A consistent and accurate ab initio parametrization of density functional dispersion correction (DFT-D) for the 94 elements H-Pu. J Chem Phys 132:154104–154119. doi:10.1063/1.3382344

95. Tkatchenko A, Scheffler M (2009) Accurate molecular van der Waals interactions from ground-state electron density and free-atom reference data. Phys Rev Lett 102:073005. doi:10.1103/PhysRevLett.102.073005

96. Becke AD (2005) Real-space post-Hartree-Fock correlation models. J Chem Phys 122:064101. doi:10.1063/1.1844493

97. Becke AD, Johnson ER (2005) A density-functional model of the dispersion interaction. J Chem Phys 123:154101. doi:10.1063/1.2065267

98. Dion M, Rydberg H, Schröder E et al (2004) Van der Waals density functional for general geometries. Phys Rev Lett 92:246401. doi:10.1103/PhysRevLett.92.246401

99. Klimeš J, Michaelides A (2012) Perspective: Advances and challenges in treating van der Waals dispersion forces in density functional theory. J Chem Phys 137:120901. doi:10.1063/1.4754130

100. Benighaus T, DiStasio RA, Lochan RC et al (2008) Semiempirical double-hybrid density functional with improved description of long-range correlation. J Phys Chem A 112:2702–2712. doi:10.1021/jp710439w

101. Takatani T, Sherrill CD (2007) Performance of spin-component-scaled Møller-Plesset theory (SCS-MP2) for potential energy curves of noncovalent interactions. Phys Chem Chem Phys 9:6106–6114. doi:10.1039/b709669k

102. Vázquez-Mayagoitia Á, Sherrill CD, Aprà E, Sumpter BG (2010) An assessment of density functional methods for potential energy curves of nonbonded interactions: The XYG3 and B97-D approximations. J Chem Theory Comput 6:727–734. doi:10.1021/ct900551z

103. Becke AD (1996) Density-functional thermochemistry. 4. A new dynamical correlation functional and implications for exact-exchange mixing. J Chem Phys 104:1040–1046. doi:10.1063/1.470829

Chapter 4
XYG3 Results for Some Selected Applications

Abstract In this chapter, some selected applications of the XYG3 functionals are described. In Sect. 4.1, a set of gas-phase reactions relevant to the Fischer–Tropsch synthesis has been constructed. With this set, we have tested the validity of the widely used PBE and B3LYP functionals, as well as XYG3. As gas-phase reactions and the corresponding surface reactions are related through the Born–Haber cycle, we argued that computational catalysis on surfaces will be less meaningful if gas-phase behaviors cannot first be suitably determined. In Sect. 4.2, we predict the heat formation of 5-chloromethylfurfural (CMF), which has been proposed as a central intermediate in the conversion of carbohydrate-based material into useful organic commodities. Using XYG3, the conversion from CMF to 5-Hydroxymethylfurfural (HMF) and levulinic acid (LA) in water, and that to biofuels 5-ethoxymethyl furfural (EMF) or ethyllevulinate (EL) in alcohol have been studied. New reaction mechanisms have been proposed, which complement the well-recognized Horvat's mechanisms. In Sect. 4.3, we have reported the XYG3 results on the processes for D-glucose pyrolysis to acrolein. It has been shown that the most feasible reaction pathway starts from an isomerization from D-glucose to D-fructose, which then undergoes a cyclic Grob fragmentation, followed by a concerted electrocyclic dehydration to yield acrolein. This study provides the first mechanism based on theory that can account for the known experimental results. In Sect. 4.4, a non-fitting protein–ligand interaction scoring function has been introduced and applied to the screening of kinase inhibitors. A good correlation has been found between the calculated scores and the experimental inhibitor efficacies with the square of correlation coefficient R^2 of 0.88 when XYG3 is used to calculate the relative binding enthalpies in the gas phase. Such a good performance can only be achieved after proper treatment of the solvation effects, as well as the entropic effects on the relative binding affinities. This represents the first high-level theory based non-fitting scoring function.

Keywords XYG3 · Gas-phase thermodynamics · Catalysis · 5-Chloromethyl-furfural · 5-Hydroxymethylfurfural · Levulinic acid · 5-ethoxymethyl furfural · Ethyllevulinate · Pyrolysis of D-glucose · Non-fitting scoring function · Kinase inhibitor

I. Y. Zhang and X. Xu, *A New-Generation Density Functional*,
SpringerBriefs in Molecular Science, DOI: 10.1007/978-3-642-40421-4_4,
© The Author(s) 2014

4.1 Gas-Phase Thermodynamics as a Prevalidation for Computational Catalysis on Surfaces

Understanding the mechanism of a catalytic process is a prerequisite for the rational design of catalysts. Even with significant advances in in situ surface chemistry techniques [1], the mechanism is still more a matter of opinion than a matter of experimental fact. There is no doubt that DFT has become the main tool in predicting the catalytic mechanisms [2]. Unfortunately, the utilization of an approximate function may have artificially biased against or favored certain mechanisms [3]. One shall be cautious about such errors and find a way to provide an early warning before going into detailed calculations on surface reactions.

Due to the scarcity of clean and reliable experimental data on surfaces, the theoretical methods used in studying heterogeneous catalytic mechanisms are usually less well validated than those for the gas-phase reactions. It was argued that gas-phase reactions and the corresponding surface reactions are related through the Born–Haber cycle and computational catalysis on surfaces will be less meaningful if gas-phase behaviors cannot first be suitably determined [4]. As schematically shown in Fig. 4.1, if admitting an error of ~ 10 kcal/mol in the gas-phase reaction energies, one will have to assume a similar amount, but with the opposite sign, of the net error for the adsorption energies to ensure a good description of the corresponding surface reactions. Such a coincidence would be too good to be true.

There has been renewed interest in the Fischer–Tropsch synthesis (FTS) due to volatile and diminishing oil supply. The mechanism of FTS is, however, a matter of ongoing debate since its discovery in the 1920s [e.g., 6–9]. Based on the available gas-phase experimental thermodynamics, [10–12] a set of gas-phase reactions have been constructed [4], which are related to FTS (see Tables 4.1, 4.2, 4.3, 4.4, 4.5). It includes C–O direct and hydrogen-assisted dissociations, hydrogenation and C–C bond couplings, as well as CH_2 and CO insertion reactions. This FTS set has been used as a testing set to check the validity of the widely used functionals (e.g., PBE and B3LYP) in their capability to explore FTS mechanisms

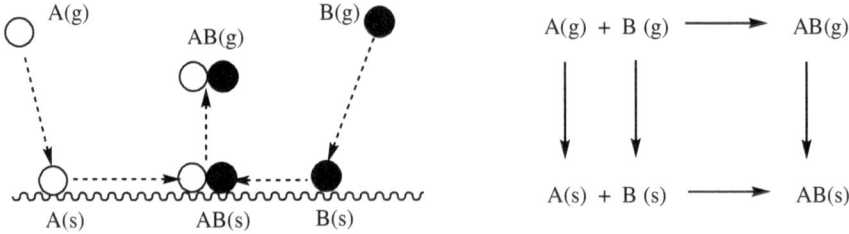

Fig. 4.1 A schematic representation of the A + B → AB reaction according to the Langmuir–Hinshelwood mechanism (*left*) and the Born–Haber cycle (*right*) that relates the gas-phase reaction to the surface reaction [4, 5]. Reprinted from Ref. [4]. Copyright 2012, with permission from John Wiley & Sons Ltd

[4]. Note that many of the reactions examined here are also important ingredients in other processes including hydrocarbon synthesis and refining, [13] hydrocarbon oxidation and combustion [13, 14] as well as CO_2 recycling [14, 15], etc.

FTS initiates through CO dissociation. However, whether it is a direct CO dissociation or hydrogen-assisted CO dissociation remains the topic of much debate [6–8]. As shown in Table 4.1, the free CO molecule possesses a strong C–O bond of 255 kcal/mol. Its strength is reduced dramatically with successive hydrogenation. Hence it can be expected that hydrogenation lowers the activation barrier heights for C–O cleavage. However, Table 4.1 suggests that PBE overestimates C–O BDEs (i.e., negative errors), which infers that C–O dissociation barriers may have been overestimated based on the Brønsted–Evans–Polanyi relationship [16]. In particular, barrier associated with HC–O dissociation may be artificially too high (by ~ 6 kcal/mol) as compared to C–O direct dissociation.

It is very likely that CO can do insertion as is common in organometallic chemistry [17]. The first two reactions listed in Table 4.2 may serve as prototypes of CO insertion into metal-H and metal-alkyl bonds. The third one in Table 4.2 may be related to the mechanism of the so-called water–gas shift reaction [7] concurrently occurs under most FTS conditions. From Table 4.2, it is clear that PBE overestimates the exothermicity of CO insertion reactions in the gas phase by various amounts (6–11 kcal/mol). If PBE can predict reliable adsorption energies, it can be expected that the tendency of CO insertions on the surfaces should also be overemphasized by PBE. As compared data in Table 4.1 and those in Table 4.2, it is clear that a method (such as represented by PBE here) can lead to negative errors for one type of reactions or positive errors for the other type of reactions, such that the mechanism can be artificially (or falsely) biased by the chosen method.

Statistically, B3LYP/XYG3 lead to MADs of 1.73/1.09 kcal/mol for C–O dissociation, and 1.13/0.76 for CO insertion, being superior to PBE in terms of gas-phase reaction energies [4].

Hydrogenation is involved in many steps during FTS. In the initial steps, it helps CO activation or does hydrogenation of surface carbide to form active CH_x species. In the chain growth steps, it transforms unsaturated surface hydrocarbons to surface alkyls, facilitates α-olefin readsorption on the surface, or does reduction of the acyl group to the alkyl group. In the chain termination steps, it releases

Table 4.1 C–O dissociation energies (kcal/mol): experimental data[a] and theoretical errors[b]

	Expt	PBE	B3LYP	XYG3
$CO \rightarrow C + O$	255.07	−10.16	3.35	0.52
$HCO \rightarrow CH + O$	191.26	−16.28	−0.81	−1.23
$H_2CO \rightarrow CH_2 + O$	178.42	−8.63	1.35	0.95
$CH_3O \rightarrow CH_3 + O$	89.66	−10.75	−1.42	1.66
MAD		11.45	1.73	1.09

Reprinted from Ref. [4]. Copyright 2012, with permission from John Wiley & Sons Ltd
[a] Experimental data are taken from Refs. [10–12]
[b] Theoretical errors are calculated via (Expt − Theor)

Table 4.2 CO insertion energies (kcal/mol): experimental data[a] and theoretical errors[b]

	Expt	PBE	B3LYP	XYG3
H–H + CO → H$_2$CO	0.46	7.53	0.34	0.62
CH$_3$–H + CO → CH$_3$CHO	4.62	8.27	−0.93	−0.10
H–OH + CO → HCOOH	−6.28	10.90	3.96	1.79
CH$_3$–OH + CO → CH$_3$COOH	−28.98	8.43	0.56	0.16
CH$_3$–OCH$_3$ + CO → CH$_3$COOCH$_3$	−27.98	7.81	0.23	0.36
CH$_3$–CH$_3$ + CO → CH$_3$COCH$_3$	−5.43	7.83	0.96	−0.32
CH$_3$–CH$_2$CH$_3$ + CO → CH$_3$COCH$_2$CH$_3$	−5.68	7.61	0.70	−0.28
CH$_3$–CCH + CO → CH$_3$COCCH	−2.18	5.99	−1.32	−2.47
MAD		8.05	1.13	0.76

Reprinted from Ref. [4]. Copyright 2012, with permission from John Wiley & Sons Ltd
[a] Experimental data are taken from Refs. [10–12]
[b] Theoretical errors are calculated via (Expt − Theor)

Table 4.3 Hydrogenation energies (kcal/mol): experimental data[a] and theoretical errors[b]

	Expt	PBE	B3LYP	XYG3
C + H → CH	−79.02	0.77	1.49	−0.62
CH + H → CH$_2$	−100.43	3.46	0.11	0.65
CH$_2$ + H → CH$_3$	−110.33	−1.74	0.40	0.46
CH$_3$ + H → CH$_4$	−104.53	−2.55	−1.83	−0.80
CO + H → HCO	−15.21	6.90	5.01	1.13
HCO + H → H$_2$CO	−87.59	−4.19	−2.50	−1.52
H$_2$CO + H → H$_2$COH	−29.75	−0.23	0.83	−0.80
H$_2$CO + H → CH$_3$O	−21.57	0.38	2.72	−0.26
CH$_3$O + H → CH$_3$OH	−103.73	−5.38	−5.10	−1.31
CH$_2$OH + H → CH$_3$OH	−95.55	−4.77	−3.21	−0.76
CH$_2$CO + H → CH$_3$CO	−42.68	−3.58	−1.41	−0.08
CH$_3$CO + H → CH$_3$CHO	−88.93	−4.69	−3.31	−1.73
CH$_3$CHO + H → CH$_3$CH$_2$O	−15.63	−0.03	2.13	−0.08
CH$_3$CH$_2$O + H → CH$_3$CH$_2$OH	−104.14	−6.55	−6.04	−1.64
CH$_3$CH$_2$ + H → CH$_3$CH$_3$	−100.61	−4.05	−2.89	−0.66
CH$_3$CHCH$_3$ + H → CH$_3$CH$_2$CH$_3$	−98.13	−5.78	−4.34	−1.00
CH$_2$ = CH$_2$ + H → CH$_3$CH$_2$	−35.27	0.48	1.59	1.22
CH$_3$CH = CH$_2$ + H → CH$_3$CHCH$_3$	−34.91	1.10	1.90	1.38
(CH$_3$)$_2$C = CH$_2$ + H → (CH$_3$)$_2$CCH$_3$	−35.33	0.93	1.36	0.61
O + H → OH	−101.22	3.24	1.54	−1.42
OH + H → H$_2$O	−118.83	−1.72	−3.78	−2.28
MAD		2.98	2.55	0.97

Reprinted from Ref. [4]. Copyright 2012, with permission from John Wiley & Sons Ltd
[a] Experimental data are taken from Refs. [10–12]
[b] Theoretical errors are calculated via (Expt − Theor)

paraffins or oxygenates. It helps to remove surface oxygen to form water. These steps are mimicked by gas-phase reactions as shown in Table 4.3.

Statistically, both PBE and B3LYP work reasonably well with MADs of 2.98 and 2.55 kcal/mol, respectively. Nevertheless, the error ranges for these functionals are still too wide. PBE errors spread from 6.90 to −6.55 kcal/mol, while those for B3LYP are within (5.01, −6.04). Hence, there are error bars larger than 10 kcal/mol associated with these two methods. For this set of reactions, it can be seen that some of the reaction enthalpies are overestimated with some others being underestimated, showing an artificial bias for some types of surface reactions. While XYG3 fares best for these hydrogenation reactions with MAD of 0.97 kcal/mol. Its error range (1.38, −2.28) is, however, not yet quite satisfactory, leaving room for further development.

An important step in FTS is the C–C coupling, which competes with the methanation step, leading to long-chain hydrocarbons. The issue of which way is the most preferred C–C coupling pathway is still under hot debate. The monomer building block CH_x was often believed to be CH_2, the mechanism can then be classified according to the primer chain C_yH_z. If C_yH_z is R (i.e., C_nH_{2n+1}), this is called the alkyl mechanism [18, 19]; if C_yH_z is CHR (i.e., C_nH_{2n}), this is called the alkylidene mechanism [20]; and if C_yH_z is CH = CHR (i.e., C_nH_{2n-1}), this is called the alkenyl mechanism [6]. Instead of using CH_2 as the building block, some other C–C coupling mechanisms have also been proposed which involve CH plus R or CHR [21], as well as C plus CR [22]. Furthermore, the chain growth may follow the so-called CO insertion mechanism [23] and the hydroxy-carbene mechanism [24].

Table 4.4 C–C coupling energies (kcal/mol): experimental data[a] and theoretical errors[b]

	Expt	PBE	B3LYP	XYG3
CH + C → HCC	−177.29	10.27	1.68	0.81
CH + CH → HCCH	−230.81	7.64	−5.45	1.02
CH + CH$_2$ → CHCH$_2$	−164.6	8.22	−0.78	0.73
CH$_2$ + CH$_2$ → CH$_2$ = CH$_2$	−174.86	−0.27	−3.84	−0.94
CH$_2$ + CH$_3$ → CH$_2$CH$_3$	−99.8	1.95	−2.47	−0.17
CH$_2$ + CH$_2$CH$_3$ → CH$_2$CH$_2$CH$_3$	−98.7	−0.58	−4.61	−0.59
CH$_3$ + CH$_3$ → CH$_3$CH$_3$	−90.08	−0.37	−5.58	−1.29
CH$_3$ + CH$_2$CH$_3$ → CH$_3$CH$_2$CH$_3$	−88.9	−2.57	−7.42	−1.31
CH$_3$ + CCH → CH$_3$CCH	−125.9	2.55	−1.57	1.49
CH$_3$ + CHCH$_2$ → CH$_3$CHCH$_2$	−101.82	−2.44	−6.82	−1.49
CH$_2$ + CO → H$_2$CCO	−78.63	12.26	3.80	1.26
CH$_3$ + CO → CH$_3$CO	−10.98	10.42	2.62	0.72
CH$_3$ + CH$_3$CO → CH$_3$COCH$_3$	−84.53	−2.96	−7.19	−2.33
2 CH$_3$OH → CH$_3$CH$_2$OH + H$_2$O	−18.01	−2.39	−3.54	−1.51
MAD		4.63	4.31	1.12

Reprinted from Ref. [4]. Copyright 2012, with permission from John Wiley & Sons Ltd
[a] Experimental data are taken from Refs. [10–12]
[b] Theoretical errors are calculated via (Expt – Theor)

The prototypical gas-phase reactions for different ways of C–C coupling are summarized in Table 4.4. PBE gives quite accurate results for (CH$_2$ + CH$_2$) and (CH$_2$ + R), whereas it overestimates the tendency for (CH + C), (CH + CH), as well as (CH + CH$_2$). Coincidently, some recent DFT studies favor the latter three coupling mechanisms against the former two on the surfaces [22, 25, 26]. Possibly, errors in the gas-phase reactions are carried over into the description of the surface reactions, such that it is the PBE functional that has chosen certain reaction mechanisms against the others.

The maximum error associated with PBE is around 12 kcal/mol for this set of C–C coupling reactions listed in Table 4.4. B3LYP halves this error to ~6 kcal/mol. In terms of MAD (i.e., 4.63 and 4.31 kcal/mol for PBE and B3LYP, respectively), these two methods behave similarly. XYG3 gives a MAD of 1.12 kcal/mol, with the largest error being −2.33 kcal/mol.

A widely favored pathway in the literature is the so-called alkyl mechanism, in which the chain growth proceeds via the insertion of the CH$_2$ species. Reactions in Table 4.5 mimic this pathway. There would be a tendency of error accumulation with CH$_2$ insertion. This is clearly seen in the results of B3LYP. PBE and XYG3 that perform well. Errors could be smaller when B3LYP is applied to study the corresponding surface reactions. Again, this relies on the error cancelation from the CH$_2$ adsorption energy on the surfaces.

Note that a good behavior of a functional in the gas-phase reactions is not necessarily a guarantee for its good description of surface reactions. It is emphasized that it is hardly conceivable that a meaningful description of a surface reaction is obtainable if the corresponding gas-phase reaction cannot first be suitably determined. Examining the related gas-phase reactions can provide important

Table 4.5 C–C chain growth according to the alkyl mechanism: experimental data[a] and theoretical errors[b]

	Expt	PBE	B3LYP	XYG3
CH$_3$–H + CH$_2$ → CH$_3$CH$_2$–H	−95.88	0.44	−3.36	−0.03
CH$_3$CH$_2$–H + CH$_2$ → CH$_3$(CH$_2$)$_2$-H	−98.62	−0.26	−4.30	−0.19
CH$_3$CH$_2$–H + 2 CH$_2$ → CH$_3$(CH$_2$)$_3$–H	−197.32	−0.66	−8.77	−0.42
CH$_3$CH$_2$–H + 3 CH$_2$ → CH$_3$(CH$_2$)$_4$–H	−296.12	−1.28	−13.45	−0.83
CH$_3$CH$_2$–H + 4 CH$_2$ → CH$_3$(CH$_2$)$_5$–H	−394.62	−1.53	−17.77	−0.86
CH$_3$CH$_2$–H + 5 CH$_2$ → CH$_3$(CH$_2$)$_6$–H	−493.32	−2.06	−22.36	−1.15
CH$_3$CH$_2$–H + 6 CH$_2$ → CH$_3$(CH$_2$)$_7$–H	−592.02	−2.51	−26.88	−1.39
MAD		1.25	13.84	0.70

Reprinted from Ref. [4]. Copyright 2012, with permission from John Wiley & Sons Ltd
[a] Experimental data are taken from Refs. [10–12]
[b] Theoretical errors are calculated via (Expt – Theor)

information on the validity of the theoretical methods for surface catalysis, hence, gas-phase thermodynamics can be and shall be used as a prevalidation for computational catalysis on surfaces.

4.2 Thermochemistry for Conversion of 5-Chloromethylfurfural into Valuable Chemicals

There is a growing interest in developing efficient biorefinery technologies for converting biomass into biofuels and valuable chemicals due to the pressing energy demands and environmental concerns. Recently, a new but possibly very valuable biomass-derived platform, 5-chloromethylfurfural (CMF) has been reported, which can be converted not only into biofuels 5-ethoxymethyl furfural (EMF) or ethyllevulinate (EL) in alcohol, but also into other value-added biomass platform such as 5-hydroxymethylfurfural (HMF) or levulinic acid (LA) in water (cf. Fig. 4.2).

However, no basic thermodynamic properties of CMF are available. In particular, no accurate energetics has been reported for the hydrolysis or alcoholysis of CMF. Such information is vitally important in addressing the challenge for thermal or catalytic conversion of biomass to fuels and useful chemicals.

First, a methodology assessment using experimental HOFs (298 K, 1 atm) [29] for selected sets of 18 furan-derivatives and 20 chloro compounds has been carried out [27]. Figures 4.3 and 4.4 display the error distributions. B3LYP is certainly not suitable for HOF calculations, whose MAD is as high as 9.4/11.7 kcal/mol for furan/chloro compounds. B3LYP consistently underestimates the stability of the species, as indicated by negative Max of $-23.5/-30.7$ kcal/mol. The same is true for B2PLYP with the present basis set of $6\text{-}311 + G(3df,2p)$. Increasing the basis set size to quadruple zeta shall reduce the errors by stabilizing the molecule more

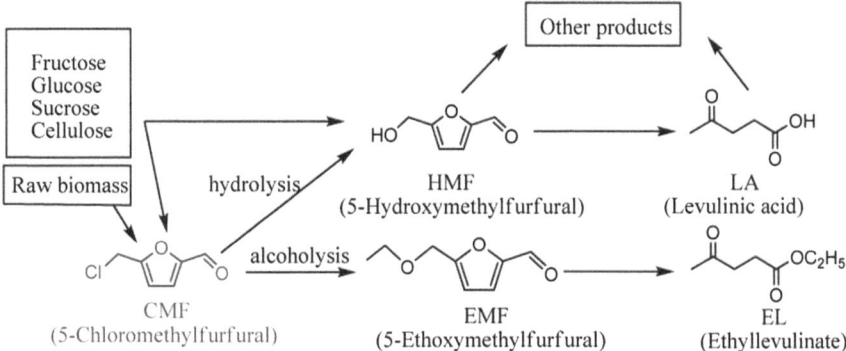

Fig. 4.2 Production of CMF and its conversion to EMF and EL, as well as its connection to other biomass platforms HMF and LA [27, 28]

than its constituent atoms. Nevertheless, this will increase dramatically the computational costs for larger molecules. On the other hand, there is a clear tendency for M06-2X to overestimate the stability of each species. MAD associated with M06-2X is 4.4/3.7 kcal/mol with positive Max of 10.0/13.8 kcal/mol for furan/ chloro compounds. G4 [30], XYG3 and X1 methods give the best performance for HOF calculations of these sets of compounds. As compared to the experimental values, they all present an MAD around 1.6 kcal/mol for the furan-derivatives and 1.9–2.8 for the chloro compounds. Note that X1 is a neural network correction method on top of B3LYP [31]. Max associated with X1, XYG3, and G4 are 7.3/ 6.5, −4.7/−8.2, 5.6/−6.9 kcal/mol, respectively, for furan/chloro compounds, which are less than half of those for B2PLYP and B3LYP in magnitude.

It should be pointed out that the experimental data may also suffer from large errors. Max (5.6 kcal/mol) for the G4 method occurs at dihydro-5-hexyl-2-furanone in the furan set. Such a large error is suspicious. The corresponding X1, M06-2X, and XYG3 HOFs are −125.9, −123.8, and −121.7 kcal/mol, respectively, lending support to the G4 value (−124.2 kcal/mol), rather than the experimental data (−118.6 kcal/mol) [29]. Similarly, we suspect the reliability of the experimental data (−106.1 kcal/mol) for chloroacetic-acid-methyl-ester $ClCH_2COOCH_3$ in the chloro set [29]. G4, X1 and XYG3 methods predict the HOF of −99.2, −100.9, and −97.9 kcal/mol, respectively for this molecule, which differ from the experimental data by more than 6 kcal/mol.

Experimental HOFs for furan-derivatives are scarce. HOF for **HMF** was measured to be −79.9 kcal/mol [32]. The corresponding G4, X1, and XYG3 values are −81.8, −80.8, and −79.2 kcal/mol, respectively, confirming the experimental data. There is no experimental HOF being reported for **CMF**. The predicted values by G4 and XYG3 are −51.2 and −50.7 kcal/mol, respectively.

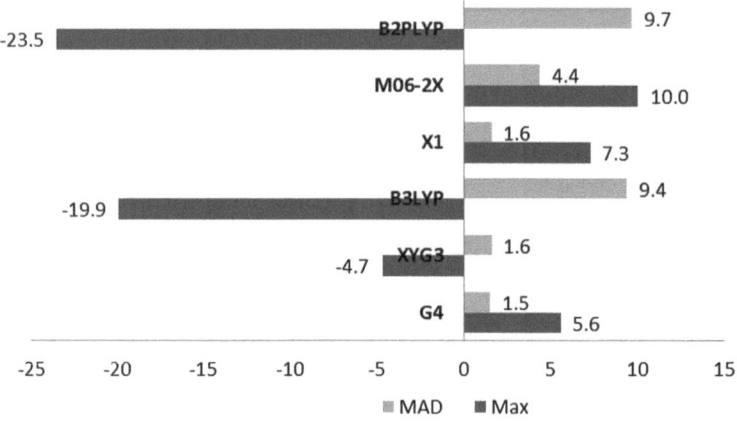

Fig. 4.3 Calculated deviations for 18 furan-derivatives at the levels of G4 theory and several DFT methods

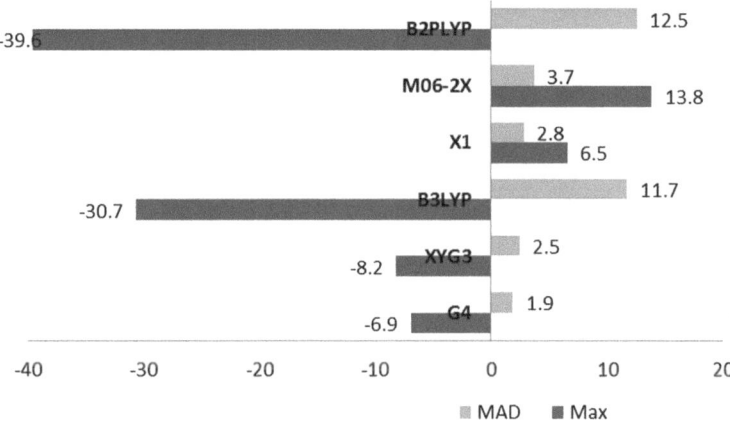

Fig. 4.4 Calculated deviations for 20 chloro compounds at the levels of G4 theory and several DFT methods

The process for conversion of CMF into LA was proposed as shown in Fig. 4.5 [27], while thermodynamic information for the detailed mechanism, at the level of XYG3, involved in the HMF to LA conversion is provided in Fig. 4.6 [27]. Solvation effects on the reaction free energy changes are investigated using the polarizable continuum model in conjunction with united atom topological model with atomic radii optimized at Hartree–Fock level (PCM-UAHF) [34].

The transformation starts by converting CMF to HMF, which is then followed by sequential HMF hydrolysis to LA. Based on the intermediates revealed by ^{13}C-NMR spectroscopy, the mechanism of the HMF to LA has been proposed by Horvat et al. [33], which is labeled as Route 1. It was proposed that HMF undergoes 4,5-addition (*R2* in Fig. 4.5) to intermediate 1 (*Int 1*). This step imposes a substantial free energy barrier due to the entropy penalty and loss of π-electron conjugation. After overcoming this barrier (16.1 kcal/mol), the reaction proceeds smoothly and goes all the way downhill. Dehydration (*R3*) from *Int 1* leads to *Int 2*, where the π-electron conjugation reappears with the formation of the butadiene moiety. Rehydration of *Int 2* (*R4*) occurs via a 1,4-like addition of butadiene by water, which is accompanied by -2.8 kcal/mol exothermicity to form *Int 3*.

The reaction comes to a fork at *Int 3*. We will first follow Route 1 as proposed by Horvat [33]. *Int 3* undergoes tautomerization to form *Int 4*, where furan ring is opened (*R5*). *R5* is favored in terms of free energy by 5.3 kcal/mol. Dehydration of *Int 4* to produce *Int 5* is slightly downhill by 0.6 kcal/mol (*R6*), possibly due to the disruption of the hydrogen bond network, but it is favored in terms of free energy as water is released to increase the entropy contribution. Originally, Horvat et al. [33] proposed that *Int 5* reacts with two water molecules to form *Int 7* and formic acid, as indicated by the dashed arrow in Fig. 4.5. We decompose it stepwise,

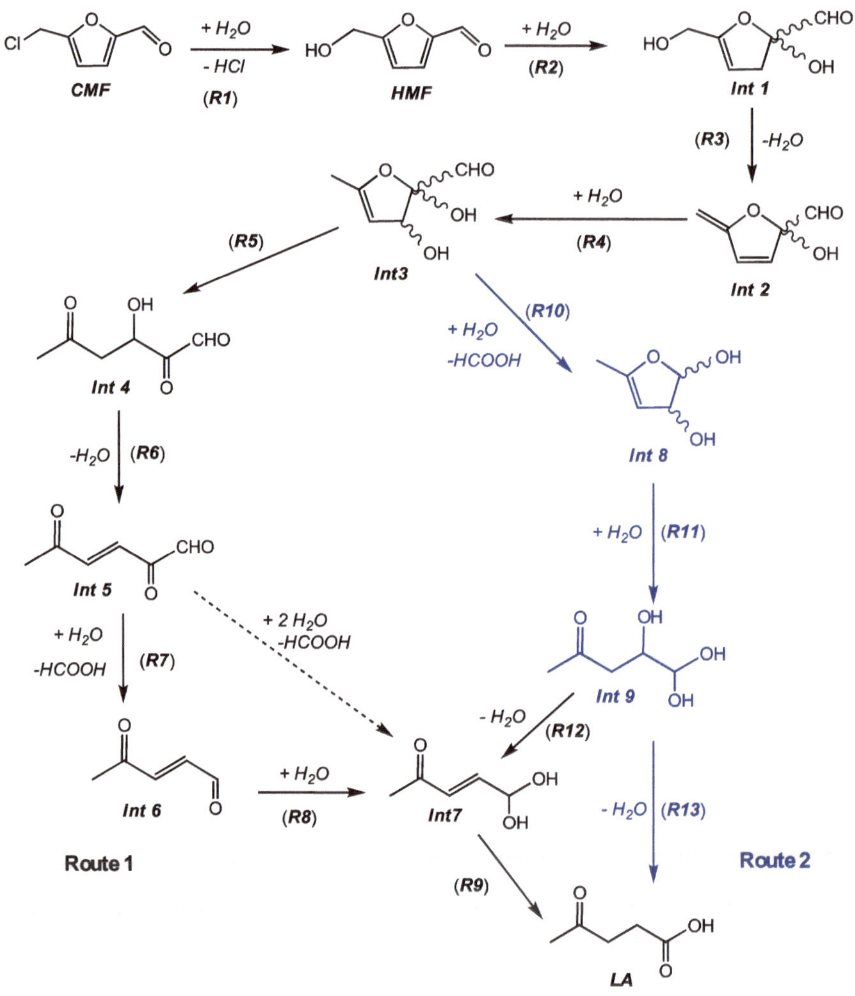

Fig. 4.5 Reaction mechanisms for the conversion of CMF into LA. Route 1 follows Horvat's original proposal [27, 33]

where **Int 5** first reacts with one water molecule to form formic acid plus **Int 6**, which then reacts with the second water to form **Int 7**. The first step (**R7**) is downhill by −13.2 kcal/mol; while the second step (**R8**) is thermoneutral. Indeed **Int 7** is just a transient, which undergoes tautomerization, leading to the final product **LA**.

We then check Route 2, which is proposed based on the XYG3 calculations as an alternative to Horvat mechanism [33]. In this route (Route 2), hydration to

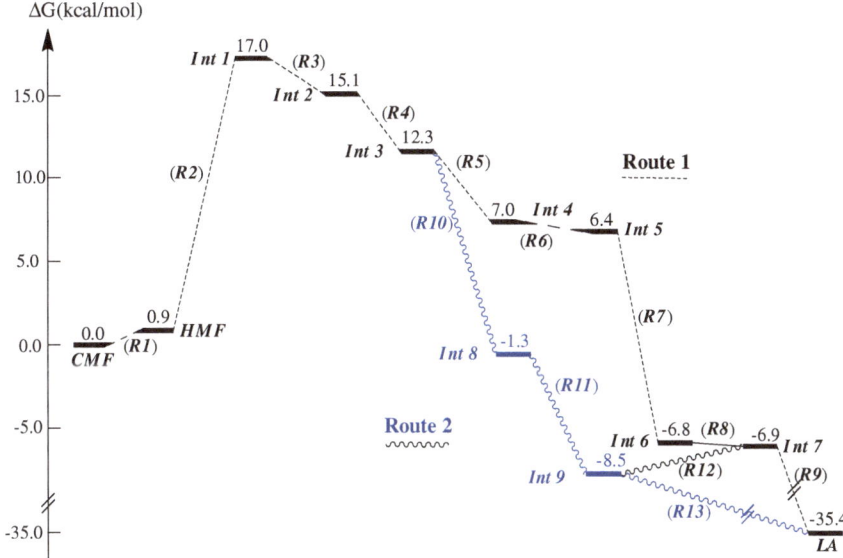

Fig. 4.6 Calculated free energy changes for the CMF conversions to LA in solution phase. Free energy changes in gas phase are calculated using XYG3 [27]. The solvation free energy changes are calculated with the PCM-UAHF model [27, 34]

release formic acid occurs first (**R10**), which is then followed by furan ring opening (**R11**). It would not be necessary to undergo dehydration to form **Int 7** via **R12**. Instead, dehydration via **R13** from **Int 7** leads to direct formation of **LA**. The large exothermicity in terms of free energy provides a strong thermodynamic driving force along this route to form **LA**.

Horvat mechanism (Route 1) is a widely recognized mechanism for **HMF** conversion to **LA**. The present calculations using XYG3 plus the PCM-UAHF model suggest an alternative pathway (Route 2) which is more thermodynamically favored (see Fig. 4.6). More detailed study is required to characterize the kinetics of Routes 1 and 2.

In analogy to the mechanisms for the hydrolysis of **CMF** to produce **HMF** and **LA**, the reaction mechanisms for the conversion of **CMF** into **EMF** and **EL** was proposed as shown in Fig. 4.7 and the calculated free energies are reproduced in Fig. 4.8. Route 3 is similar to Route 1 in that de-alcoholization happens first, which is then complemented by decarboxylation and alcoholization; while Route 4 is similar to Route 2 which undergoes decarboxylation and alcoholization first, followed by de-alcoholization. In contrast to hydrolysis where Route 2 is much more favored than Route 1, Route 3, and Route 4 were found to be comparable in terms of free energy changes.

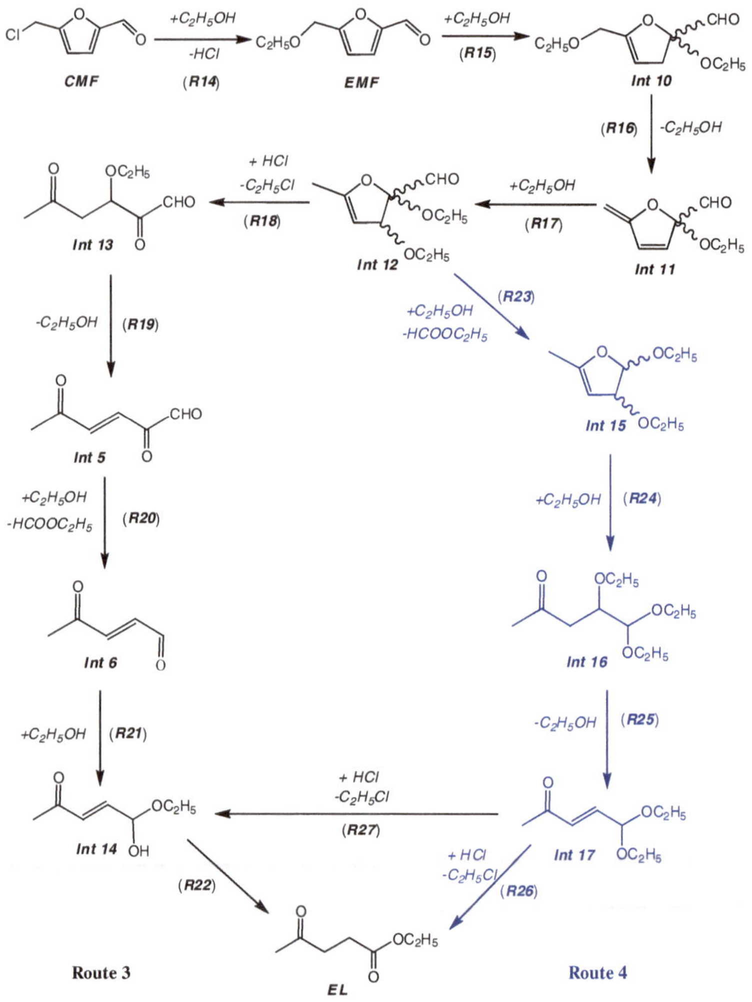

Fig. 4.7 Proposed reaction mechanism for the conversion of CMF to EL [27]

4.3 Reaction Mechanisms for Pyrolysis of D-glucose to Acrolein

Understanding of the chemistry of D-glucose pyrolysis is vitally important, which acts as a useful representative of carbohydrate pyrolysis, being heavily involved not only in the utilization of biomass [35, 36], but also in food industry and tobacco industry [36, 37]. Simple aldehydes, such as formaldehyde, acetaldehyde, and acrolein, are formed during pyrolysis [38–40]. These simple aldehydes are not only widely used industrial chemicals, but also classified as carcinogens.

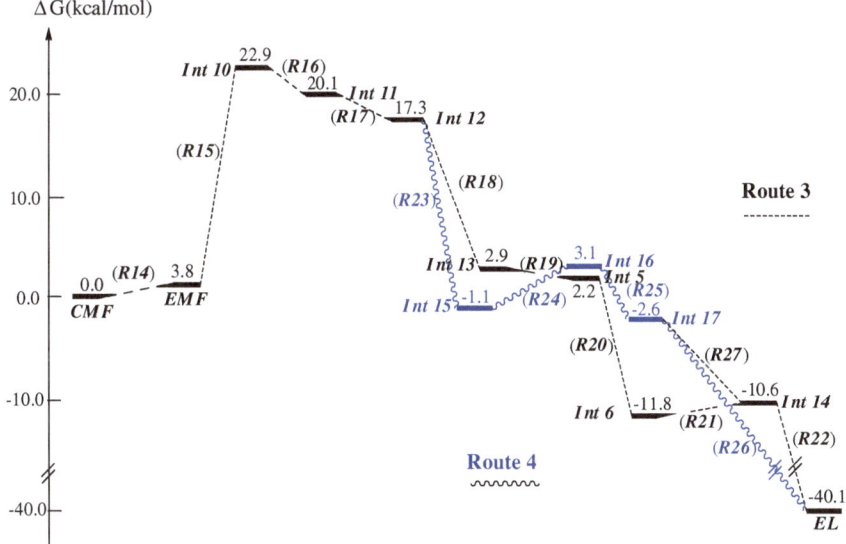

Fig. 4.8 Calculated free energy changes for the CMF conversions to EL in solution phase. Free energy changes in gas phase are calculated using XYG3 [27]. The solvation free energy changes are calculated with the PCM-UAHF model [34]

In spite of this importance, much of the detailed mechanism of carbohydrate pyrolysis is still unknown [38–43], even though a recent systematical experimental work of Paine et al. [38–40] has partially filled the gap. Using variously labeled ^{13}C in conjunction with gas chromatography/mass spectroscopy (GC/MS), possible mechanisms of D-glucose pyrolysis have been suggested, although no theoretical calculations have been carried out to quantify the mechanisms, partially because the calculations of reaction barrier heights are complicated by the network of hydrogen bonds.

In the literature, there are some related theoretical studies. For instance, Abella et al. [43] have investigated the reaction mechanisms of cellulose pyrolysis and levoglucosan decomposition by using B3LYP as well as MP2 [43]. Nimols et al. [42] have performed CBS-QB3 [44] calculations to investigate dehydration mechanisms of neutral glycerol. As a specific validation of XYG3 for carbohydrate chemistry, the B3LYP and XYG3 energetics for glycerol dehydration have been first checked. As compared to the accurate CBS-QB3 results, MAD associated with XYG3/6-311 + G(3df,2p) is 2.2 kcal/mol, while the corresponding MAD of B3LYP with the same basis set is 7.7 kcal/mol.

The first theoretical study on the pyrolysis mechanisms from D-glucose to acrolein was reported. Geometry optimizations and frequency calculations were carried out by using B3LYP/6-311G(d,p). The final energies were evaluated by single point calculations using XYG3/6-311 + G(3df,2p). All energies were reported as ΔG in kcal/mol at 700 K to be in line with the experimental condition [36].

It is expected that D-glucose shall bear a heavy conformational complexity due to the formation of internal hydrogen bonds. To simplify the calculations, the open chain conformer (1 in Fig. 4.9) was chosen as the initial reactant for each reaction

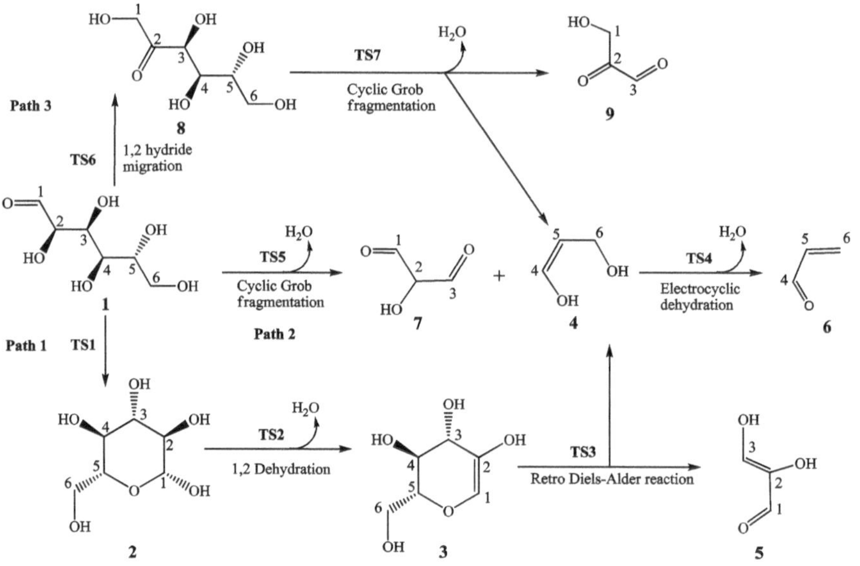

Fig. 4.9 Reaction mechanisms for pyrolysis of D-glucose to acrolein

path. The terminal aldehyde carbon is labeled as C-1, and all six carbons are labeled sequentially. The possible reaction mechanisms for pyrolysis of D-glucose to acrolein are summarized in Fig. 4.9. The calculated reaction energetics are depicted in Fig. 4.10, and the optimized transition state structures are shown in Fig. 4.11.

Path 1 starts by forming a hemiacetal **2** at C-5, which is an anomer of six-membered pyranose, being 5.9 kcal/mol more stable than glucose **1** in the open chain form. From **1** to **2**, a barrier of 40.0 kcal/mol (**TS1**) has to be overcome. This is followed by a 1,2-dehydration process, where a hydroxyl group at C-1 and a hydrogen at C-2 are expelled as a water to generate a cyclic enol **3**. This is the rate determining step of **Path 1**, as the associated barrier height (**TS2**) is as high as 67.9 kcal/mol. Fragmentation is carried out via a retro-Diels–Alder reaction. The corresponding reaction barrier was found to be 44.5 kcal/mol (**TS3**). Fragment **4** (prop-1-ene-1,3-diol) is the alkene and fragment **5** (2,3-dihydroxy acrylaldehyde) is the corresponding diene in the Diels–Alder reaction. Fragment **4** is made of C-4, C-5 and C-6, which may undergo a concerted electrocyclic dehydration with a barrier height of 27.7 kcal/mol (**TS4**), leading to the target product acrolein **6**.

Cyclic Grob fragmentation was believed to be the main reaction path for carbohydrate degradation [38–40]. A barrier of 63.1 kcal/mol was found for such a process from **1** via **TS5** (**Path 2**). Proton in the hydroxyl group at C-3 would be transferred to the hydroxyl group at C-5, losing the 5-oxygen as water. Meanwhile, the C–C bond between C-3/C-4 would be cleaved, forming fragments **4** and **7**. Fragment **7** (2-hydroxy malonaldehyde) is a tautomer of fragment **5** in Path 1, while fragment **4** would again undergo electrocyclic dehydration, yielding the desired product acrolein **6**.

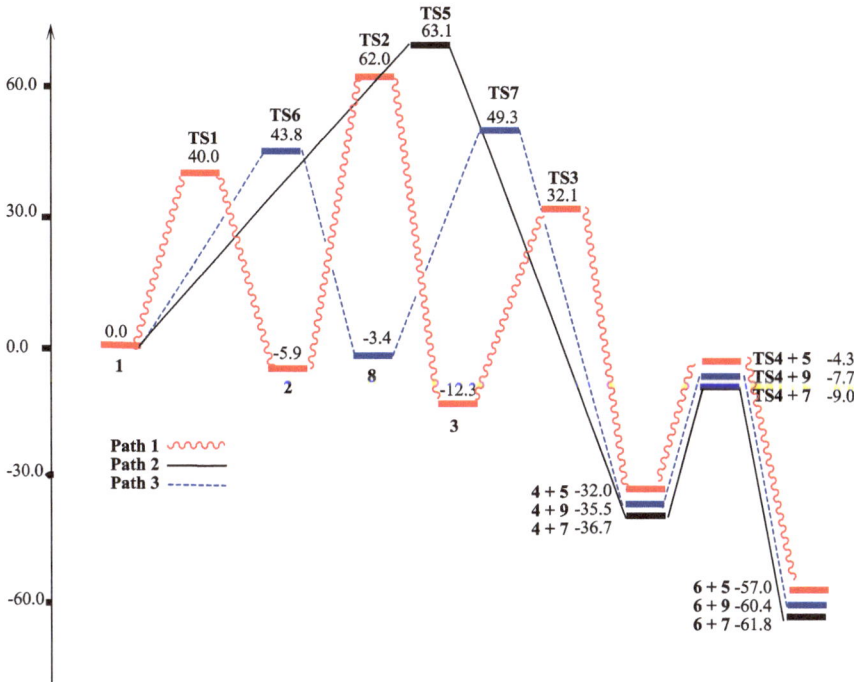

Fig. 4.10 Schematic potential energy curves for pyrolysis of D-glucose to acrolein. Energies are in kcal/mol

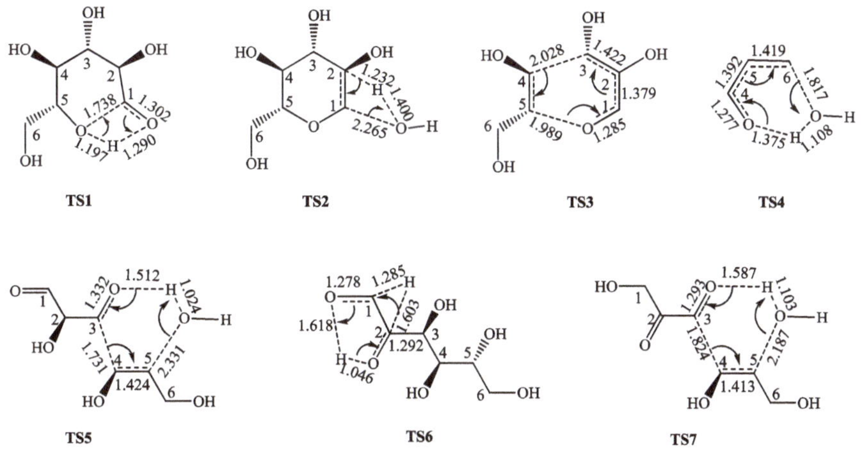

Fig. 4.11 The optimized geometries of the transition states for pyrolysis of D-glucose to acrolein. Bond distances are in Å

XYG3 calculations suggest that the isomerization of D-glucose to D-fructose can also be realized in gas-phase pyrolysis without going through enolization as a base-catalyzed reaction. As shown in **Path 3** in Fig. 4.9, a one-step process can be initiated via a 1,2 hydride migration that exchanges the carbonyl and the hydroxyl group between C-1 and C-2. This converts D-glucose **1** to D-fructose **8**. The corresponding transition state (**TS6**) is shown in Fig. 4.11, where the hydride at C-2 attacks the electro-positive C-1, and simultaneously, proton of hydroxyl at C-2 is transferred to the terminal aldehyde oxygen. The activation barrier for this step was calculated to be 43.8 kcal/mol. The next step is the cyclic Grob fragmentation of D-fructose **8** via **TS7**, producing fragments **4** and **9**. As compared to that via **TS5**, barrier height via **TS7** is lowered by 10.3 kcal/mol, which may be attributed to a more favorable structure for internal hydrogen bondings. Fragment **9** (3-hydroxy-2-oxo propanal) is an isomer of **5** in **Path 1** and **7** in **Path 2**. It was found that **9** and **7** are 3.5 and 4.7 kcal/mol, respectively, more stable than 5. As compared to the other two pathways, **Path 3** is the most favorable. The barrier from **8** to **TS7** is 52.8 kcal/mol.

There is an earlier experimental conjecture of Stein's for the pyrolysis mechanism of D-glucose. Radicals were assumed to be formed via a homolytic cleavage of C–C bond between C-1 and C-2 [41]. And an activation energy (55.6 ± 5.5 kcal/mol) was reported [41]. Paine criticized Stein's conjecture as being not consistent with the energy needed for C–C bond homolysis [38–40]. Instead, by means of isotopic ^{13}C labeling and GC/MS, Paine et al. proposed the alternative pathways [38–40]. As 63 % of all the acrolein formed was found to be from C-4, C-5, and C-6, with the aldehyde group derived specially from C-4 and the vinyl group derived from C-5 and C-6, they suggested that the dominant mechanisms should be unimolecular, where there were two competing channels as **Path 1** and **Path 2** shown in Fig. 4.9 [38–40]. Nevertheless, there were no corresponding calculations reported up to now to quantify Paine's proposals [39]. The XYG3 calculations fulfill such a purpose, giving strong support to Paine's cyclic unimolecular mechanisms. It was shown that **Path 1** and **Path 2** are indeed competitive. Particularly, the best pathway is identified as **Path 3** where glucose is first converted to fructose, which then undergoes cyclic Grob fragmentation. **Path 3**, possessing an effective barrier of 49.3 kcal/mol, also yields acrolein made of C-4, C-5, and C-6, in good agreement with known experimental results.

4.4 A Non-Fitting Protein–Ligand Interaction Scoring Function Applied to the Screening of Kinase Inhibitors

Currently, a hot topic in the fields of cancer research and drug design is the so-called targeted therapy [45, 46]. For such an approach to be successful, it is necessary to develop potent and selective inhibitors for the identified target protein. An accurate theoretical prediction of the protein-inhibitor interaction is, therefore, vital to the screening and discovery of the inhibitors, which could significantly accelerate the rational design of an effective drug. However, current ways of estimating inhibitor efficacy rely on empirical protein–ligand interaction scoring functions which are usually heavily parameterized. A significant limitation lies in that one should not use an empirical method to treat a system that is beyond its parameterization range, whereas a drug candidate is quite possibly a new molecule, such that the good performance of an empirical method cannot be guaranteed. Even for the molecules that are within the parameterization range, the accuracy is limited, whose squared correlation coefficient R^2 generally ranges only from 0.1 to 0.5 [47].

In order to get improvement over empirical scoring methods, several difficulties have to be first surmounted to introduce quantum mechanical (QM) methods [48–51]. For example, as atoms involved in the short-range interactions between a drug molecule and the nearby protein residues can often add up to hundreds of atoms, the QM region has to be sufficiently large, while at the same time, the QM calculations have to be carried out efficiently. As nonbonded interactions often play an important role in protein–drug interactions, the QM methods employed have to be capable of describing nonbonded interactions accurately. Entropic and solvation effects have to be considered properly.

Recently, a non-fitting scoring function has been proposed [52], which consists of three terms:

$$\text{Score} = \Delta\Delta H^{\text{gas phase}} + \Delta\Delta G^{\text{solv}} + T\Delta\Delta S \qquad (4.1)$$

The first term is the relative protein–ligand binding enthalpy in gas phase.

$$\Delta\Delta H^{\text{gas phase}} = H^{\text{gas},0k}_{\text{protein−ref}} - H^{\text{gas},0k}_{\text{protein−targ}} - H^{\text{gas},0k}_{\text{ref}} + H^{\text{gas},0k}_{\text{targ}} \qquad (4.2)$$

$H^{\text{gas},0k}_{\text{protein−ref}}$ and $H^{\text{gas},0k}_{\text{protein−targ}}$ represent the electronic structure energies of the reference and the target protein–ligand complexes, respectively. This was obtained by using an eXtended ONIOM (XO) method [53, 54], which introduced the "divide and conquer" strategy [55] into the original ONIOM method [56] to enable a cheap and accurate high-level description of a very large QM region based on an integration of DFT methods (e.g., XYG3) and the semi-empirical PM6 [57] method.

The second term is the solvation effect on the relative binding affinity, which mainly reflects the penalty for transferring a solvated drug molecule into the binding pocket of the protein.

$$\Delta\Delta G^{\text{solv}} = \Delta G^{\text{solv}}_{\text{protein−ref}} - \Delta G^{\text{solv}}_{\text{protein−targ}} - \Delta G^{\text{solv}}_{\text{ref}} + \Delta G^{\text{solv}}_{\text{targ}} \qquad (4.3)$$

The ΔG^{solv} calculations have taken advantage of the newly developed SMD solvation model, which was parameterized to give good total solvation free energies for over a hundred solvents [58].

The third term is the entropy effect, which mainly reflects the penalty for a free drug molecule being trapped in the binding pocket of the protein. This was estimated by using standard DFT frequency analysis [52]. Since proteins are soft matters, it is reasonable to expect that there is still certain flexibility even when the drug molecule has already bound to the proteins. A simple approximation is to consider that the ligand molecule is "half trapped", loosing half of its freedom

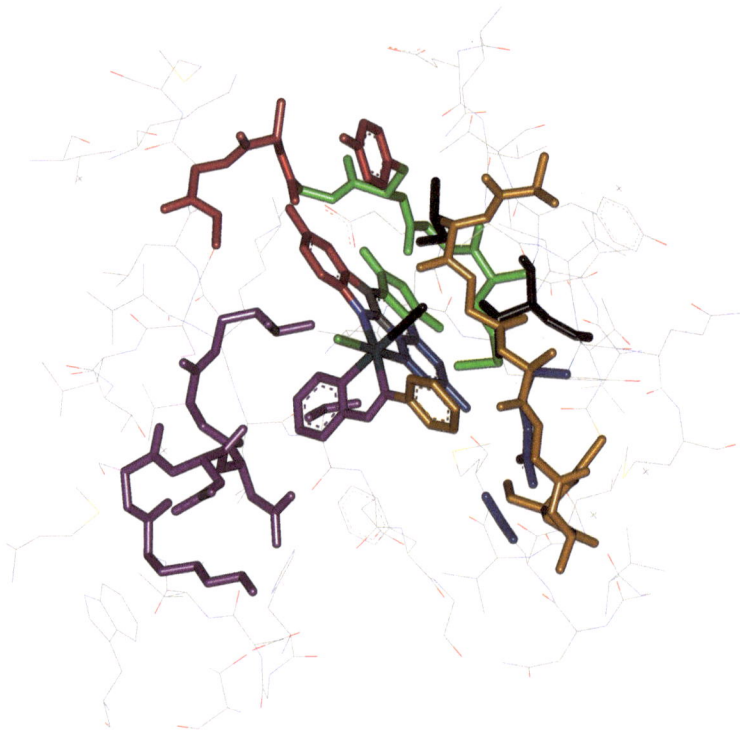

Fig. 4.12 The eXtended ONIOM (XO) scheme for the calculation of the PAK1-FL172 complex. The high-level region is shown in stick model. The high-level region is divided into seven overlapping fragments. Six of them are used to cover the nonbonded interaction between the protein residues and the functional groups of the drug molecule (shown in different colors). One of them is used to describe the whole drug molecule

upon binding. Hence 50 % of the free ligand molecule entropy is taken as the penalty on binding affinity due to entropic effect.

An example is given for the PAK1-FL172 reference complex (See Fig. 4.12). The function of protein kinas PAK1 is to regulate cell motility and morphology, and has been suggested as a target in cancer therapy [59]. A previous experimental work results in a PAK1 inhibitor database, which includes the chemically inert organometallic lead structure FL172 and its 19 derivatives, with the corresponding data for inhibited PAK1 activity [60, 61].

Table 4.6 shows the experimental data ln[Activity/(100 − Activity)], the calculated protein–ligand interaction score of 20 PAK1 inhibitors and the corresponding components. The data show that the three components actually make comparable, in many cases opposite, contributions to the final score. Figure 4.13a shows the plot of the theoretical scores (x) versus the experimental data (y) for the PAK1 inhibitor dataset. The theoretical results and experimental data display significant linear correlations. The R^2 coefficient found is as high as 0.88.

Table 4.6 Experimental data, calculated protein–ligand interaction score and its components of 20 PAK1 inhibitors (FL172 is the Ref. [52])

Ligand	$\Delta\Delta H^{a,b}$ (gas phase)	$\Delta\Delta G(Solv)^{a,c}$	$T\Delta\Delta S^{a,d}$	Score	$\ln[A/(100\text{-}A)]$
FL172	0.00	0.00	0.00	0.00	−1.52
FL237	−0.61	−1.90	−1.71	−4.23	0.58
FL252	−1.28	−0.59	−0.49	−2.36	−0.58
FL254	6.63	−5.32	−3.50	−2.19	−1.39
FL256	4.76	−4.41	−0.63	−0.27	−1.99
FL258	−0.35	−0.72	−3.49	−4.57	−0.12
FL2901	0.17	−1.71	0.00	−1.54	−1.15
FL293	−0.19	−0.70	−1.55	−2.44	−0.36
FL408	1.51	−0.96	−2.70	−2.14	−0.94
FL410	−0.46	−0.17	−2.78	−3.40	−0.75
FL327	−1.58	1.48	−1.19	−1.29	−1.27
FL343	−1.15	−4.27	−1.18	−6.60	1.45
FL363	1.32	−1.08	−1.82	−1.58	−1.59
FL411	6.76	−4.87	−0.88	1.00	−1.82
FL735	1.09	−0.59	−1.70	−1.19	−1.66
FL752	4.69	−2.52	−2.40	−0.23	−2.02
FL809	8.94	−5.01	−3.08	0.85	−1.82
FL1088	2.28	−1.22	−0.78	0.28	−2.70
FL07111	−4.81	−0.91	1.11	−4.61	−0.08
FL134	−6.49	1.37	−1.20	−6.31	1.32

[a] Unit: kcal/mol
[b] $\Delta\Delta H$(gas phase) $= \Delta\Delta H^{gas\ phase}$ is the gas phase relative binding enthalpy given by the eXtended ONIOM (XO) calculations using XYG3 as the highest level
[c] $\Delta\Delta G(Solv) = \Delta\Delta G^{Solv}$ is the solvation effect on the relative binding affinity given by SMD cluster calculations
[d] $T\Delta\Delta S$ is the entropic effect on the relative binding affinity given by ωB97-D frequency analysis

Figure 4.13b, c displays the plots for the correlations between the calculated components and the experimental data of inhibitor efficacies. In fact, none of the components alone shows a good correlation with the experimental data. The R^2 coefficient from $\Delta\Delta H^{gas\ phase}$ is 0.48, whereas those from $\Delta\Delta G^{solv}$ and $T\Delta\Delta S$ are only 0.06 and 0.003, respectively. This demonstrates that all of the components have to be accurately evaluated and combined together to give a satisfactory description of the protein–ligand interactions [52].

It should be emphasized that a better method should always lead to a better performance for a first-principles based score function. This is in sharp contrast to the heavily parameterized empirical scoring functions, where improvements on the calculations of certain descriptors usually lead to a declined performance, unless the whole scoring function is re-parameterized. Furthermore, the non-fitting scoring function should be applied well to different proteins, while a significant degradation in accuracy is always observed in the trans-protein applications of empirical scoring functions. The same score function has been successfully testified on a CDK2 inhibitor database including 76 CDK2 protein inhibitors [52, 62].

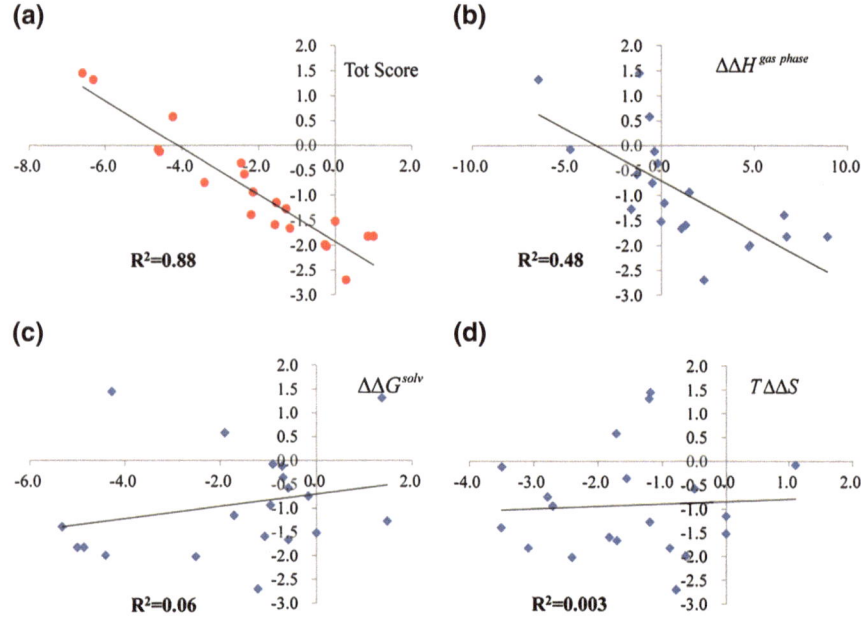

Fig. 4.13 a Plot of the theoretical scores (*x*) versus the experimental data ln[Activity/(100 −
Activity)] (*y*) for the 20 PAK1-inhibitor complexes. Theoretical scores are obtained by using
XYG3. The points above the fitted trend-line indicate an overestimated protein–ligand interaction
by the score function, while the points under the fitted trend-line indicate an underestimation. Plot
of the calculated components (*x*) versus the experimental data (*y*). The components are: **b** gas-
phase relative binding enthalpy $\Delta\Delta H^{gas\,phase}$ obtained by XYG3; **c** solvation effect $\Delta\Delta G^{solv}$;
d entropic effect [52]

References

1. Somorjai GA, Li Y (2011) Impact of surface chemistry. Proc Natl Acad Sci USA
   108:917–924. doi:10.1073/pnas.1006669107
2. Nørskov JK, Abild-Pedersen F, Studt F, Bligaard T (2011) Density functional theory in
   surface chemistry and catalysis. Proc Natl Acad Sci USA 108:937–943. doi:10.1073/
   pnas.1006652108
3. Jónsson H (2011) Simulation of surface processes. Proc Natl Acad Sci USA 108:944–949.
   doi:10.1073/pnas.1006670108
4. Zhang IY, Xu X (2012) Gas-phase thermodynamics as a validation of computational catalysis
   on surfaces: a case study of Fischer-Tropsch synthesis. ChemPhysChem 13:1486–1494.
   doi:10.1002/cphc.201100909
5. Masel RI (1996) Principles of adsorption and reaction on solid surfaces. John Wiley, New
   York, p 444
6. Maitlis PM, Zanotti V (2009) The role of electrophilic species in the Fischer-Tropsch
   reaction. Chem Commun 1619–1634. doi:10.1039/b822320n
7. Van der Laan GP, Beenackers AACM (1999) Kinetics and selectivity of the Fischer–Tropsch
   synthesis: a literature review. Catal Rev Sci Eng 41:255–318. doi:10.1081/CR-100101170

8. Davis BH (2009) Fischer–Tropsch synthesis: reaction mechanisms for iron catalysts. Catal Today 141:25–33. doi:10.1016/j.cattod.2008.03.005
9. Hindermann JP, Hutchings GJ, Kiennemann A (1993) Mechanistic aspects of the formation of hydrocarbons and alcohols from CO hydrogenation. Catal Rev Sci Eng 35:1–127. doi:10.1080/01614949308013907
10. Curtiss LA, Raghavachari K, Redfern PC, Pople JA (2000) Assessment of Gaussian-3 and density functional theories for a larger experimental test set. J Chem Phys 112:7374–7383. doi:10.1063/1.481336
11. Curtiss LA, Raghavachari K, Redfern PC, Pople JA (1997) Assessment of Gaussian-2 and density functional theories for the computation of enthalpies of formation. J Chem Phys 106:1063–1079. doi:10.1063/1.473182
12. Wu JM, Xu X (2008) Improving the B3LYP bond energies by using the X1 method. J Chem Phys 129:164103–164111. doi:10.1063/1.2998231
13. Olah GA, Molnár Á (2003) Hydrocarbon chemistry, 2nd edn. John Wiley, New Jersey
14. Inderwildi OR, Jenkins SJ (2008) In-silico investigations in heterogeneous catalysis-combustion and synthesis of small alkanes. Chem Soc Rev 37:2274–2309. doi:10.1039/b719149a
15. Huber GW, Iborra S, Corma A (2006) Synthesis of transportation fuels from biomass: chemistry, catalysts, and engineering. Chem Rev 106:4044–4098. doi:10.1021/cr068360d
16. Van Santen RA, Neurock M, Shetty SG (2010) Reactivity theory of transition-metal surfaces: a Bronsted-Evans-Polanyi linear activation energy-free-energy analysis. Chem Rev 110:2005–2048. doi:10.1021/cr9001808
17. Keim W (ed) (1983) Catalysis in C1 chemistry. D. Reidel Publishing Company, Dordrecht
18. Biloen P, Helle JN, Sachtler WMH (1979) Incorporation of surface carbon into hydrocarbons during Fischer-Tropsch synthesis: mechanistic implications. J Catal 58:95–107. doi:10.1016/0021-9517(79)90248-3
19. Brady RC, Pettit R (1981) Mechanism of the Fischer-Tropsch reaction. The chain propagation step. J Am Chem Soc 103:1287–1289. doi:10.1021/ja00395a081
20. Dry ME (1996) Practical and theoretical aspects of the catalytic Fischer-Tropsch process. Appl Catal A-Gen 138:319–344. doi:10.1016/0926-860X(95)00306-1
21. Ciobica IM, Kramer GJ, Ge Q et al (2002) Mechanisms for chain growth in Fischer-Tropsch synthesis over Ru(0001). J Catal 212:136–144. doi:10.1006/jcat.2002.3742
22. Liu ZP, Hu P (2002) A new insight into Fischer-Tropsch synthesis. J Am Chem Soc 124:11568–11569. doi:10.1021/ja012759w
23. Schulz H (2010) Advances in Fischer-Tropsch synthesis, catalysts, and catalysis. CRC Press, Taylor & Francis Group, Florida, p 165
24. Kummer JT, Emmett PH (1953) Fischer—Tropsch synthesis mechanism studies. The addition of radioactive alcohols to the synthesis gas. J Am Chem Soc 75:5177–5183. doi:10.1021/ja01117a008
25. Chen J, Liu ZP (2008) Origin of selectivity switch in Fischer – Tropsch synthesis over Ru and Rh from first-principles statistical mechanics studies. J Am Chem Soc 130:7929–7937. doi:10.1021/ja7112239
26. Cheng J, Hu P, Ellis P et al (2010) Some understanding of Fischer-Tropsch synthesis from density functional theory calculations. Top Catal 53:326–337. doi:10.1007/s11244-010-9450-7
27. Liu G, Wu J, Zhang IY et al (2011) Theoretical studies on thermochemistry for conversion of 5-chloromethylfurfural into valuable chemicals. J Phys Chem A 115:13628–13641. doi:10.1021/jp207641g
28. Mascal M, Nikitin EB (2008) Direct, high-yield conversion of cellulose into biofuel. Angew Chem-Int Edit 47:7924–7926. doi:10.1002/anie.200801594
29. NIST standard reference database number 69 (2011) http://webbook.nist.gov/chemistry Accessed 15 Aprl 2013
30. Curtiss L, Redfern P, Raghavachari K (2007) Gaussian-4 theory. J Chem Phys 126:084108–084112. doi:10.1063/1.2436888

31. Wu JM, Xu X (2007) The X1 method for accurate and efficient prediction of heats of formation. J Chem Phys 127:214105–214113. doi:10.1063/1.2800018
32. Verevkin SP, Emel'yanenko VN, Stepurko EN et al (2009) Biomass-derived platform chemicals: thermodynamic studies on the conversion of 5-hydroxymethylfurfural into bulk intermediates. Ind Eng Chem Res 48:10087–10093. doi:10.1021/ie901012g
33. Horvat J, Klaić B, Metelko B, Šunjić V (1985) Mechanism of levulinic acid formation. Tetrahedron Lett 26:2111–2114. doi:10.1016/S0040-4039(00)94793-2
34. Barone V, Cossi M, Tomasi J (1997) A new definition of cavities for the computation of solvation free energies by the polarizable continuum model. J Chem Phys 107:3210–3221. doi:10.1063/1.474671
35. Mohan D, Pittman CU, Steele PH (2006) Pyrolysis of wood/biomass for bio-oil: a critical review. Energy Fuels 20:848–889. doi:10.1021/ef0502397
36. Baker RR, Coburn S, Liu C (2006) The pyrolytic formation of formaldehyde from sugars and tobacco. J Anal Appl Pyrolysis 77:12–21. doi:10.1016/j.jaap.2005.12.009
37. Talhout R, Opperhuizen A, van Amsterdam JGC (2006) Sugars as tobacco ingredient: effects on mainstream smoke composition. Food Chem Toxicol 44:1789–1798. doi:10.1016/j.fct.2006.06.016
38. Paine JB, Pithawalla YB, Naworal JD (2008) Carbohydrate pyrolysis mechanisms from isotopic labeling. Part 2. The pyrolysis of D-glucose: general disconnective analysis and the formation of C-1 and C-2 carbonyl compounds by electrocyclic fragmentation mechanisms. J Anal Appl Pyrolysis 82:10–41. doi:10.1016/j.jaap.2008.01.002
39. Paine JB, Pithawalla YB, Naworal JD (2008) Carbohydrate pyrolysis mechanisms from isotopic labeling. Part 3. The pyrolysis of D-glucose: formation of C-3 and C-4 carbonyl compounds and a cyclopentenedione isomer by electrocyclic fragmentation mechanisms. J Anal Appl Pyrolysis 82:42–69. doi:10.1016/j.jaap.2007.12.005
40. Paine JB, Pithawalla YB, Naworal JD (2008) Carbohydrate pyrolysis mechanisms from isotopic labeling Part 4. The pyrolysis of D-glucose: the formation of furans. J Anal Appl Pyrolysis 83:37–63. doi:10.1016/j.jaap.2008.05.008
41. Stein YS, Antal MJ Jr, Jones M Jr (1983) A study of the gas-phase pyrolysis of glycerol. J Anal Appl Pyrolysis 4:283–296. doi:10.1016/0165-2370(83)80003-5
42. Nimlos MR, Blanksby SJ, Qian XH et al (2006) Mechanisms of glycerol dehydration. J Phys Chem A 110:6145–6156. doi:10.1021/jp060597q
43. Abella L, Nanbu S, Fukuda K (2007) Memoirs of the Faculty of Engineering. Kyushu University 67:67
44. Montgomery JA, Frisch MJ, Ochterski JW, Petersson GA (1999) A complete basis set model chemistry. VI. Use of density functional geometries and frequencies. J Chem Phys 110:2822–2827. doi:10.1063/1.477924
45. Zhang J, Yang PL, Gray NS (2009) Targeting cancer with small molecule kinase inhibitors. Nat Rev Cancer 9:28–39. doi:10.1038/nrc2559
46. Wlodawer A, Vondrasek J (1998) Inhibitors of HIV-1 protease: a major success of structure-assisted drug design. Annu Rev Biophys Biomolec Struct 27:249–284. doi:10.1146/annurev.biophys.27.1.249
47. Li SY, Xi LL, Wang CQ et al (2009) A novel method for protein-ligand binding affinity prediction and the related descriptors exploration. J Comput Chem 30:900–909. doi:10.1002/jcc.21078
48. Raha K, Merz KM (2004) A quantum mechanics-based scoring function: study of zinc ion-mediated ligand binding. J Am Chem Soc 126:1020–1021. doi:10.1021/ja038496i
49. Grater F, Schwarzl SM, Dejaegere A et al (2005) Protein/ligand binding free energies calculated with quantum mechanics/molecular mechanics. J Phys Chem B 109:10474–10483. doi:10.1021/jp044185y
50. Raha K, Peters MB, Wang B et al (2007) The role of quantum mechanics in structure-based drug design. Drug Discov Today 12:725–731. doi:10.1016/j.drudis.2007.07.006

51. Hayik SA, Dunbrack R, Merz KM (2010) Mixed quantum mechanics/molecular mechanics scoring function to predict protein-ligand binding affinity. J Chem Theory Comput 6:3079–3091. doi:10.1021/ct100315g

52. Rao L, Zhang IY, Guo W et al (2013) Nonfitting protein–ligand interaction scoring function based on first-principles theoretical chemistry methods: Development and application on kinase inhibitors. J Comput Chem 34:1636–1646. doi:10.1002/jcc.23303

53. Guo WP, Wu AA, Xu X (2010) XO: An extended ONIOM method for accurate and efficient geometry optimization of large molecules. Chem Phys Lett 498:203–208. doi:10.1016/j.cplett.2010.08.033

54. Guo WP, Wu AA, Zhang IY, Xu X (2012) XO: an extended ONIOM method for accurate and efficient modeling of large systems. J Comput Chem 33:2142–2160. doi:10.1002/jcc.23051

55. Yang WT (1991) Direct calculation of electron density in density-functional theory. Phys Rev Lett 66:1438–1441. doi:10.1103/PhysRevLett.66.1438

56. Svensson M, Humbel S, Froese RDJ et al (1996) ONIOM: a multilayered integrated MO + MM method for geometry optimizations and single point energy predictions. A test for Diels-Alder reactions and $Pt(P(t\text{-}Bu)_3)_2 + H_2$ oxidative addition. J Phys Chem 100:19357–19363. doi:10.1021/jp962071j

57. Stewart JJP (2007) Optimization of parameters for semiempirical methods V. Modification of NDDO approximations and application to 70 elements. J Mol Model 13:1173–1213. doi:10.1007/s00894-007-0233-4

58. Marenich AV, Cramer CJ, Truhlar DG (2009) Universal solvation model based on solute electron density and on a continuum model of the solvent defined by the bulk dielectric constant and atomic surface tensions. J Phys Chem B 113:6378–6396. doi:10.1021/jp810292n

59. Kumar R, Gururaj AE, Barnes CJ (2006) P21-activated kinases in cancer. Nat Rev Cancer 6:459–471. doi:10.1038/nrc1892

60. Maksimoska J, Feng L, Harms K et al (2008) Targeting large kinase active site with rigid, bulky octahedral ruthenium complexes. J Am Chem Soc 130:15764–15765. doi:10.1021/ja805555a

61. Feng L, Geisselbrecht Y, Blanck S et al (2011) Structurally sophisticated octahedral metal complexes as highly selective protein kinase inhibitors. J Am Chem Soc 133:5976–5986. doi:10.1021/ja1112996

62. Alzate-Morales J, Caballero J (2010) Computational study of the interactions between guanine derivatives and cyclin-dependent kinase 2 (CDK2) by CoMFA and QM/MM. J Chem Inf Model 50:110–122. doi:10.1021/ci9003027

Chapter 5
Concluding Remarks

Abstract In the application of Kohn-Sham density functional theory (KS-DFT), the exchange-correlation energy must be approximated. A ladder of such approximations has been proposed, none of which is equally good for every problem. There is still a long way to go. In this chapter, we first give a brief summary of what we have learned in pursuing an improved functional Sect. 5.1, giving a list of the doubly hybrid density functionals (DHDFs) developed till date in the literature. We then outline, in Sect. 5.2, the limitations and the anticipated future development for the XYG3 type of DHDFs. Finally, a perspective is presented, which highlights some fundamental issues in the ground state KS-DFT.

Keywords Density functional theory · Exchange-correlation · XYG3 · Doubly hybrid density functionals

5.1 Lessons Learned

Kohn-Sham density functional theory (KS-DFT) replaces the correlated wave-function problem by a more tractable problem of non-interacting electron system. Although exact in principle, KS-DFT requires in practice an approximation to the exchange-correlation functional. With increasingly sophisticated approximations, KS-DFT has now become the most widely used method for electronic structure calculations, and has made great contribution to our understanding of molecular science.

This book focuses on some recent advances in construction of the so-called doubly hybrid density functionals (DHDFs). It is our opinion that DHDFs currently available shall be classified into three groups according to how they are constructed based on the underlying principles, or technically which orbitals are used to evaluate the second-order perturbative correlations.

I. Y. Zhang and X. Xu, *A New-Generation Density Functional*,
SpringerBriefs in Molecular Science, DOI: 10.1007/978-3-642-40421-4_5,
© The Author(s) 2014

Table 5.1 Summary of the MC3BB type of DHDFs

Name	DFT exchange + correlation	Compounds of wavefunction methods	Ref.
MC3BB	B88 + B95	HF/MP2	[1]
MC3MPW	mPW + PW91	HF/MP2	[1]
MC3MPWB	mPW + B95	HF/MP2	[2]
MC3TS	TPSS + KCIS	HF/MP2	[2]
MCCO-MPW	mPW + PW91	HF/MP2	[2]
MCCO-MPWB	mPW + B95	HF/MP2	[2]
MCCO-TS	TPSS + KCIS	HF/MP2	[2]
MCUT-MPW	mPW + PW91	HF/MP2/MP4(SDQ)	[2]
MCUT-MPWB	mPW + B95	HF/MP2/MP4(SDQ)	[2]
MCUT-TS	TPSS + KCIS	HF/MP2/MP4(SDQ)	[2]
MCQCISD-MPW	mPW + PW91	HF/MP2/QCISD	[2]
MCQCISD-MPWB	mPW + B95	HF/MP2/QCISD	[2]
MCQCISD-TS	TPSS + KCIS	HF/MP2/QCISD	[2]
MCG3-MPW	mPW + PW91	HF/MP2/MP4(SDQ)/QCISD(T)	[2]
MCG3-MPWB	mPW + B95	HF/MP2/MP4(SDQ)/QCISD(T)	[2]
MCG3-TS	TPSS + KCIS	HF/MP2/MP4(SDQ)/QCISD(T)	[2]

Table 5.1 lists the first type of DHDFs, which mix the total energies of a DFT calculation and those of wavefunction methods [1, 2]. The latter are not limited to MP2 (Møller-Plesset perturbation theory at second order) as in MC3BB [1], but updated even to the QCISD(T) level [2] (i.e., quadratic configuration interaction with single and double plus fourth-order and fifth-order quasi-perturbative terms involving triple excitations). They are actually multi-coefficient methods. One can think of these methods as improving the correlation part of hybrid DFT, or one can think of them as adding static correlation and additional dynamic correlation to the best practical single reference wavefunction theory methods. It has been argued that, in the eye of KS-DFT, the exchange-correlation energy in this class of DHDFs depends on both Hartree-Fock (HF) orbitals and KS orbitals, both of which are functionals of the density [3].

The second type is represented by B2PLYP [4]. Technically, it is very similar to the standard MP2. Instead of using HF orbitals, it uses orbitals generated from a hybrid DFT that already contains partial correlation. Such a hybrid DFT is not supposed to be used alone, which has to be completed after the PT2 term is augmented. Later, the theoretical basis of the B2PLYP type of DHDFs is provided by the multi-determinant extension of the Kohn-Sham scheme [5]. Table 5.2 summarizes the DHDFs in this class [4–18].

We proposed the third type of DHDFs to use orbitals from a fully functionalized DFT [19]. The key idea of the XYG3 type of functionals is to combine the Görling-Levy (GL) coupling-constant perturbation theory [20] and the standard KS scheme [21] in the framework of the adiabatic connection formalism [22, 23]. In particular, XYG3 can be considered as a natural evolution of the well-tested

Table 5.2 Summary of the B2PLYP type of DHDFs

Name	Description	Ref.
B2PLYP	B88 exchange; LYP correlation; PT2 correlation based on hybrid-GGA part	[4]
mPW2PLYP	Like B2PLYP, but with mPW exchange	[6]
B2KPLYP	Re-parameterized B2PLYP version for kinetics	[7, 8]
B2TPLYP	Re-parameterized B2PLYP version for thermochemistry	[7, 8]
B2GPPLYP	Re-parameterized B2PLYP version for general purpose applications	[8]
B2πPLYP	Re-parameterized B2PLYP version for conjugated π-systems	[9]
B2P3LYP	Modified B2PLYP version with long-range PT2 correction	[10]
B2OS3LYP	Similar to B2P3LYP, but with SOS-PT2 correlation	[10]
ROB2PLYP	Re-parameterized B2PLYP version within a restricted KS formalism for treating open-shell systems	[11]
ωB97X-2	Ingredients of the B97 functional; long-range corrected; SCS-PT2 correlation	[12]
B2-PPW91	Like B2PLYP, but with PW91 correlation	[13]
DSD-BLYP	Modified B2PLYP version with SCS-PT2 correction; fitted together with DFT-D dispersion correction	[14]
DSD-PBEP86	Like DSD-PBEP86, but with PBE exchange and P86 correlation	[15]
1DH-BLYP	One-parameter version of B2PLYP	[5]
PTPSS-D3	Re-optimized TPSS exchange and correlation; SOS-PT2 correlation; fitted together with DFT-D3 dispersion correction	[16]
PWPB95-D3	Re-optimized PW exchange and B95 correlation; SOS-PT2 correlation; fitted together with DFT-D3 dispersion correction	[16]
PBE0-DH	No fitting parameters based on PBE exchange–correlation	[17]
PBE0-2	No fitting parameters based on PBE exchange–correlation	[18]

hybrid functional B3LYP [24, 25]. Because of using B3LYP orbitals, XYG3 shares with B3LYP the kinetic energy, the classic Coulomb interaction energies between electrons and electrons, as well as those between electrons and nuclei. These are the larger terms in the total energy of any interested system. As compared to B3LYP, only the exchange-correlation term is updated in XYG3, where each term, including the PT2 contribution, is evaluated using the B3LYP orbitals. Other fully functionalized standard KS functionals can serve the same role as B3LYP in XYG3, while the parameters in the DHDFs shall be re-parameterized accordingly, as, after all, we only have an approximate KS functional at hand to generate the orbitals and the GL theory is truncated at the second order. Table 5.3 summarizes the DHDFs in this class [19, 26–30].

There are several intensive tests for DHDFs of different kinds (e.g., See Refs. [17, 31–33]). Our observation is that all these DHDFs contain fitting parameters which can be tuned, leading to similar accuracies, and are especially useful for the main group chemistry. Not every functional is equally applicable to every problem. This makes benchmarking evitable in choosing the right functional for the right problem.

Table 5.3 Summary of the XYG3 type of DHDFs

Name	Description	Ref.
XYG3	B88 exchange; LYP correlation; evaluated with B3LYP orbitals and densities	[19]
XYG3s	Like XYG3, but introducing the scaling all correlation (SAC) method to speed the PT2 evaluation	[26]
XYG3o	Like XYG3, but with re-optimized parameters for smaller basis sets	[27]
XYGJ-OS	Like XYG3, but with SOS-PT2 correlation	[28]
xDH-PBE0	PBE exchange and correlation; evaluated with PBE0 orbitals and densities; SOS-PT2 correlation	[29]
lrc-XYG3	XYG3 plus scaled long-range PT2 correlation	[30]

5.2 Limitations

There are limitations in the current version of the XYG3 family of functionals, which point to the direction of future development.

(1) The functionals are trained and validated with benchmark sets developed for the main group chemistry. Although some DHDFs, such as B2PLYP [7, 8, 14–16], are shown to have encouraging results for some transition metal involved complexes, systematic validation and possible extension are needed for the XYG3 family of functionals being applied confidently to transition metal chemistry.

(2) Only single-point energy calculations have been carried out for the results of the XYG3 type of DHDFs presented in this book, which may fail if the B3LYP geometries diverge significantly from the experimental geometries. Potential energy scans suggest that XYG3 should be able to give accurate geometries for either covalently bonded or noncovalently bonded complexes with standard basis sets commonly used [31], more systematical investigation can only be made possible when the analytical gradients are feasible. Development of analytical gradient has just appeared [34], while development of analytical hessian shall be made possible in the near future.

(3) Evaluation of the PT2 term scales formally as N^5, where N measures the system size. The SAC model has been adopted to speed evaluation of the PT2 terms in DHDFs [26]. XYGJ-OS has been implemented, which explores the possibility of using techniques of resolution of identity and scaled-opposite-spin MP2 [28]. We anticipate that other methods developed for the second-order Møller-Plesset perturbation theory can be readily used in DHDFs for efficient calculations of large molecules.

(4) Extensions to include other properties (e.g., electric properties and nuclear magnetic properties, etc.), excited states and periodic systems are valuable directions to go.

5.3 Perspective

Despite its great success, there are fundamental issues that remain to be answered in ground state density functional theory. For example, Ruzsinszky and Perdew have raised 12 outstanding problems [35]. They are

Problem 1: Problems of finding the minimum.
Problem 2: Is there a systematic way to construct density functionals?
Problem 3: Approximating the density functional for the kinetic energy.
Problem 4: What is the correct long-range behavior of the exchange–correlation hole?
Problem 5: Remaining problems of semi-local functionals.
Problem 6: Can we construct a proper density functional for non-collinear magnetism?
Problem 7: Can we make fourth-rung density functionals without empiricism?
Problem 8: What are the range-separated hybrid functionals telling us?
Problem 9: Can we describe both long-range charge transfer and static correlation?
Problem 10: Can we find a useful correction to the random phase approximation?
Problem 11: What is the best way to incorporate long-range van der Waals interactions?
Problem 12: Is symmetry breaking acceptable?

More recently, Cohen, Mori-Sánchez and Yang have reviewed the challenges for density functional theory [36]. Five challenges they summarized are:

Challenge 1: To develop a functional that performs uniformly better than B3LYP.
Challenge 2: The need to improve the description of reaction barriers and dispersion/van der Waals interactions.
Challenge 3: To understand the significance of $E[\rho]$ vs $E[\{\Phi_i, \varepsilon_i\}]$, OEP (optimized effective potential), and beyond.
Challenge 4: Delocalization error and static correlation error.

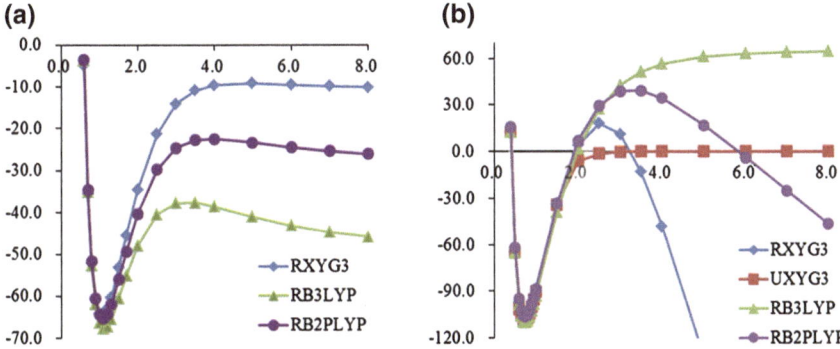

Fig. 5.1 a Dissociation curves of H_2^+; **b** Dissociation curves of H_2. All calculations are performed using the 6-311 + G(3df,2p) basis set

Challenge 5: The energy of two protons separated by infinity with one and two electrons: Strong correlation.

The basic errors of current DFT functionals, mostly from the first to the fourth rungs, have been highlighted by their abilities to describe the stretched H_2^+ and H_2 systems [36]. Figure 5.1 displays the performance of DHDFs. From Fig. 5.1 it is seen that restricted XYG3 (RXYG3) provides a good, although not perfect, description of the stretched H_2^+, but it fails for the H_2 system. If symmetry breaking is acceptable, unrestricted XYG3 (UXYG3), as well as other functionals, is able to give good description of the H_2 system.

The holy grail in KS-DFT is to find better and better approximations to the exact exchange-correlation functional. There is still a long way to go.

References

1. Zhao Y, Lynch BJ, Truhlar DG (2004) Doubly hybrid meta DFT: new multi-coefficient correlation and density functional methods for thermochemistry and thermochemical kinetics. J Phys Chem A 108:4786–4791. doi:10.1021/jp049253v
2. Zhao Y, Lynch BJ, Truhlar DG (2005) Multi-coefficient extrapolated density functional theory for thermochemistry and thermochemical kinetics. Phys Chem Chem Phys 7:43–52. doi:10.1039/b416937a
3. Zhao Y, Meana-Pañeda R, Truhlar DG (2012) MLGAUSS-version 3.0. University of Minnesota, Minneapolis
4. Grimme S (2006) Semiempirical hybrid density functional with perturbative second-order correlation. J Chem Phys 124:034108. doi:10.1063/1.2148954
5. Sharkas K, Toulouse J, Savin A (2011) Double-hybrid density-functional theory made rigorous. J Chem Phys 134:064113. doi:10.1063/1.3544215
6. Schwabe T, Grimme S (2006) Towards chemical accuracy for the thermodynamics of large molecules: new hybrid density functionals including non-local correlation effects. Phys Chem Chem Phys 8:4398 4401. doi:10.1039/b608478h
7. Tarnopolsky A, Karton A, Sertchook R et al (2008) Double-hybrid functionals for thermochemical kinetics. J Phys Chem 112:3–8. doi:10.1021/jp710179r
8. Karton A, Tarnopolsky A, Lamère J-F et al (2008) Highly accurate first-principles benchmark data sets for the parametrization and validation of density functional and other approximate methods. Derivation of a robust, generally applicable, double-hybrid functional for thermochemistry and thermochemical kinetics. J Phys Chem A 112:12868–12886. doi:10.1021/jp801805p
9. Sancho-García JC, Pérez-Jiménez AJ (2009) Assessment of double-hybrid energy functionals for pi-conjugated systems. J Chem Phys 131:084108. doi:10.1063/1.3212881
10. Benighaus T, DiStasio RA, Lochan RC et al (2008) Semiempirical double-hybrid density functional with improved description of long-range correlation. J Phys Chem A 112:2702–2712. doi:10.1021/jp710439w
11. Graham D, Menon A, Goerigk L et al (2009) Optimization and basis-set dependence of a restricted-open-shell form of B2-PLYP double-hybrid density functional theory. J Phys Chem A 113:9861–9873. doi:10.1021/jp9042864
12. Chai J-D, Head-Gordon M (2009) Long-range corrected double-hybrid density functionals. J Chem Phys 131:174105. doi:10.1063/1.3244209
13. Mohajeri A, Alipour M (2012) B2-PPW91: a promising double-hybrid density functional for the electric response properties. J Chem Phys 136:124111. doi:10.1063/1.3698284

14. Kozuch S, Gruzman D, Martin JML (2010) DSD-BLYP: a general purpose double hybrid density functional including spin component scaling and dispersion correction. J Phys Chem C 114:20801–20808. doi:10.1021/jp1070852

15. Kozuch S, Martin JML (2011) DSD-PBEP86: in search of the best double-hybrid DFT with spin-component scaled MP2 and dispersion corrections. Phys Chem Chem Phys 13:20104–20107. doi:10.1039/C1CP22592H

16. Goerigk L, Grimme S (2011) Efficient and accurate double-hybrid-meta-GGA density functionals—evaluation with the extended GMTKN30 database for general main group thermochemistry, kinetics, and noncovalent interactions. J Chem Theor Comput 7:291–309. doi:10.1021/ct100466k

17. Brémond E, Adamo C (2011) Seeking for parameter-free double-hybrid functionals: the PBE0-DH model. J Chem Phys 135:024106. doi:10.1063/1.3604569

18. Chai J-D, Mao S-P (2012) Seeking for reliable double-hybrid density functionals without fitting parameters: the PBE0-2 functional. Chem Phys Lett 538:121–125. doi:10.1016/j.cplett.2012.04.045

19. Zhang Y, Xu X, Goddard WA (2009) Doubly hybrid density functional for accurate descriptions of nonbond interactions, thermochemistry, and thermochemical kinetics. Proc Natl Acad Sci USA 106:4963–4968. doi:10.1073/pnas.0901093106

20. Görling A, Levy M (1993) Correlation-energy functional and its hight-density limit obtained from a coupling-constant perturbation expansion. Phys Rev B 47:13105–13113. doi:10.1103/PhysRevB.47.13105

21. Kohn W, Sham LJ (1965) Self-consistent equations including exchange and correlation effects. Phys Rev 140:A1133–A1138. doi:10.1103/PhysRev.140.A1133

22. Gunnarsson O, Lundqvist BI (1976) Exchange and correlation in atoms, molecules, and solids by the spin-density-functional formalism. Phys Rev B 13:4274–4298. doi:10.1103/PhysRevB.13.4274

23. Langreth DC, Perdew JP (1977) Exchange-correlation energy of a metallic surface: wave-vector analysis. Phys Rev B 15:2884–2901. doi:10.1103/PhysRevB.15.2884

24. Becke AD (1993) Density-functional thermochemistry. 3.: the role of exact exchange. J Chem Phys 98:5648–5652. doi:10.1063/1.464913

25. Stephens PJ, Devlin FJ, Chabalowski CF, Frisch MJ (1994) Ab-initio calculation of vibrational absorption and circular-dichroism spectra using density-functional force-fields. J Phys Chem 98:11623–11627. doi:10.1021/j100096a001

26. Zhang IY, Luo Y, Xu X (2010) XYG3 s: Speedup of the XYG3 fifth-rung density functional with scaling-all correlation method. J Chem Phys 132:194105. doi:10.1063/1.3424845

27. Zhang I, Luo Y, Xu X (2010) Basis set dependence of the doubly hybrid XYG3 functional. J Chem Phys 133:104105. doi:10.1063/1.3488649

28. Zhang IY, Xu X, Jung Y, Goddard WA (2011) A fast doubly hybrid density functional method close to chemical accuracy using a local opposite spin ansatz. Proc Natl Acad Sci USA 108:19896–19900. doi:10.1073/pnas.1115123108

29. Zhang IY, Su NQ, Brémond ÉAG et al (2012) Doubly hybrid density functional xDH-PBE0 from a parameter-free global hybrid model PBE0. J Chem Phys 136:174103 doi:10.1063/1.3703893

30. Zhang IY, Xu X (2013) Reaching a uniform accuracy for complex molecular systems: long-range-corrected XYG3 doubly hybrid density functional. J Phys Chem Lett 4:1669–1675. doi:10.1021/jz400695u

31. Zhang IY, Xu X (2011) Doubly hybrid density functional for accurate description of thermochemistry, thermochemical kinetics and nonbonded interactions. Int Rev Phys Chem 30:115–160. doi:10.1080/0144235X.2010.542618

32. Burns LA, Vázquez-Mayagoitia AV, Sumpter BG, Sherrill CD (2011) Density-functional approaches to noncovalent interactions: a comparison of dispersion corrections (DFT-D), exchange-hole dipole moment (XDM) theory, and specialized functionals. J Chem Phys 134:084107. doi:10.1063/1.3545971

33. Zhang IY, Xu X (2012) XYG3 and XYGJ-OS performances for noncovalent binding energies relevant to biomolecular structures. Phys Chem Chem Phys 14:12554–12570. doi:10.1039/c2cp40904f
34. Su NQ, Zhang IY, Xu X, Analytic derivatives for the XYG3 type of doubly hybrid density functionals: theory, implementation, and assessment. J Comput Chem 34:1759–1774. doi:10.1002/jcc.23312
35. Ruzsinszky A, Perdew JP (2011) Twelve outstanding problems in ground-state density functional theory: a bouquet of puzzles. Comput Theor Chem 963:2–6. doi:10.1016/j.comptc.2010.09.002
36. Cohen AJ, Mori-Sánchez P, Yang WT (2011) Challenges for density functional theory. Chem Rev 112:289–320. doi:10.1021/cr200107z